OWSO6R07b

OVERVIEW OF STATE ENERGY REDUCTION PROGRAMS AND GUIDELINES FOR THE WASTEWATER SECTOR

by:

Joseph C. Cantwell, P.E.
William R. King
Robert T. Lorand, P.E.
Science Applications International Corporation

Robert C. Ganley, P.E.
Nathyn M. Knipe, P.E.
David I. Page
O'Brien & Gere

2010

WERF

NYSERDA

New York State
Energy Research and
Development Authority

The Water Environment Research Foundation, a not-for-profit organization, funds and manages water quality research for its subscribers through a diverse public-private partnership between municipal utilities, corporations, academia, industry, and the federal government. WERF subscribers include municipal and regional water and wastewater utilities, industrial corporations, environmental engineering firms, and others that share a commitment to cost-effective water quality solutions. WERF is dedicated to advancing science and technology addressing water quality issues as they impact water resources, the atmosphere, the lands, and quality of life.

For more information, contact:
Water Environment Research Foundation
635 Slaters Lane, Suite G-110
Alexandria, VA 22314-1177
Tel: (703) 684-2470 Fax: (703) 299-0742 www.werf.org werf@werf.org

This report was co-published by the following organization.

IWA Publishing
Alliance House, 12 Caxton Street
London SW1H 0QS, United Kingdom
Tel: +44 (0) 20 7654 5500 Fax: +44 (0) 20 7654 5555 www.iwapublishing.com publications@iwap.co.uk

Library of Congress Catalog Card Number: 2009941914
IWAP ISBN: 978-1-843393-351/1-84339-335-2

Science Applications International Corporation, O'Brien & Gere

The research on which this report is based was developed, in part, by the United States Environmental Protection Agency (EPA) through Cooperative Agreement No. CR83155901-2 with the Water Environment Research Foundation (WERF). However, the views expressed in this document are not necessarily those of the EPA and EPA does not endorse any products or commercial services mentioned in this publication. This report is a publication of WERF, not EPA. Funds awarded under the Cooperative Agreement cited above were not used for editorial services, reproduction, printing, or distribution.

ACKNOWLEDGMENTS

This project was conducted as part of the WERF's Operations Optimization Challenge. Under this challenge, WERF funded projects that can assist wastewater engineers, designers, operators, or other practitioners in achieving a significant improvement in wastewater or solids treatment operations through economically and environmentally responsible process optimizations for energy, cost, and environmental footprint in a carbon-constrained world. This project was a collaborative effort between WERF and the New York State Energy Research and Development Authority (NYSERDA).

Research Team

Principal Investigator:
Joseph C. Cantwell, P.E.
Science Applications International Corporation

Project Team:
William R. King
Robert T. Lorand, P.E.
Science Applications International Corporation

Robert C. Ganley, P.E.
Nathyn M. Knipe, P.E.
David I. Page
O'Brien & Gere

Technical Review Committee

Robert K. Bastian
U.S. Environmental Protection Agency

Shahid Chaudhry
California Energy Commission

Stephen Constable, P.E.
DuPont Engineering Research & Technology

Ted Jones
Consortium for Energy Efficiency

Kathleen O'Connor, P.E.
New York State Energy Research and Development Authority (NYSERDA)

James Wheeler, P.E.
U.S. Environmental Protection Agency

Issue Area Team

John Barber, Ph.D.
Eastman Chemical

Shahid Chaudhry
California Energy Commission

Stephen Constable, P.E.
DuPont Engineering Research & Technology

David Cooley
Hampton Roads Sanitation District

Glen T. Daigger, Ph.D., P.E.
CH2M HILL

Robert F. Kelly
Infilco Degremont, Inc.

Melanie S. Masek, P.E.
DuPont Engineering Technology

Arthur J. Meyers, Jr., Ph.D.
Eastman Chemical (retired)

Hsiao-Ting Ong, P.E.
City of San Jose, CA

Ali Oskouie, Ph.D.
Metro Water Reclamation District of Greater Chicago

Steven M. Rogowski, P.E.
Metro Wastewater Reclamation District of Denver

Gary Shimp
Black & Veatch

Robert Simm, Ph.D.
Stantec Consulting, Inc.

David W. Tucker
City of San Jose, CA

Dariush Vosooghi, Ph.D., P.E.
City of Los Angeles, CA

Milind Wable, Ph.D., P.E.
CDM

James Wheeler, P.E.
U.S. Environmental Protection Agency

John Willis, P.E.
Brown & Caldwell

Water Environment Research Foundation Staff

Director of Research: Daniel M. Woltering, Ph.D.
Program Director: Lauren Fillmore, M.S.

ABSTRACT AND BENEFITS

Abstract:

Approximately 30-35% of wastewater treatment facilities' total cost of service is for energy use. The research presents two approaches for reducing facilities' energy use. Currently, energy efficiency wording is lacking in nearly all state-level design guidelines for wastewater treatment facilities; therefore, the goal of the first approach is to develop draft model design guideline wording that addresses energy efficiency. The research includes an overview of energy efficiency wording in design guidelines for the 50 states and the Federal government. It also presents model wording based on three states' wording (i.e., Pennsylvania, Virginia, and Washington) and on comments from five municipal wastewater treatment organizations. The research suggests pathways for incorporating energy efficiency into design guidelines, lists barriers to be addressed, and presents an outline for an energy best practice guidebook as a resource for selecting energy-efficient design options.

The goal of the second approach is to present a program management framework for a model wastewater system energy reduction program that state agencies and utilities can implement. The framework is based on a review of program management best practices in eight state and national programs. The review shows substantial commonality in best practices among these programs.

Benefits:

- Provides draft model energy efficiency wording as a starting point for states' efforts in revising design guidelines.
- Documents current lack of energy efficiency wording in nearly all state-level design guidelines for wastewater treatment facilities.
- Provides a program management framework for a model wastewater system energy reduction program based on the best practices of existing programs.
- Documents commonality in best practices for program management among existing wastewater treatment facility energy reduction programs.

Keywords: Energy efficiency, design guidelines, best practices, program management.

TABLE OF CONTENTS

LIST OF TABLES

LIST OF ACRONYMS

ACEEE	American Council for an Energy-Efficient Economy
ADEM	Alabama Department of Environmental Management
ADOP	Anaerobic Digester Optimization Pilot
ASHRAE	American Society of Heating, Refrigerating and Air Conditioning Engineers
AWWA	American Water Works Association
BPA	Bonneville Power Administration
CALMAC	California Measurement Advisory Council
CalPOP	California Wastewater Process Optimization Program
CEC	California Energy Commission
CEQ	Commission on Environmental Quality
CHP	Combined heat and power
CPUC	California Public Utility Commission
CWSRF	Clean Water State Revolving Fund
DEM	Department of Environmental Management
DEQ	Department of Environmental Quality
DEQ-WCRO	Department of Environmental Quality – West Central Regional Office
DHH	Department of Health and Hospitals
DNR	Department of Natural Resources
DNREC	Department of Natural Resources and Environmental Control
DO	Dissolved oxygen
EFC	New York State Environmental Facilities Corporation
EPA	Environmental Protection Agency
EPRI	Electric Power Research Institute
F.A.C.	Florida Administrative Code
GBMSD	Green Bay Metropolitan Sewage District
GELCAT	Green Energy Life Cycle Assessment Tool
GWh	gigawatt hour
IAT	Issue Area Team
ID	identification
KCWTD	King County Wastewater Treatment Division
KDHE	Kansas Department of Health and Environment
kWh	kilowatts per hour
LCAMER	Life Cycle Assessment Manager for Energy Recovery
LCCA	life cycle cost analysis

LEED	Leadership in Energy and Environmental Design (Green Building Rating System, developed by the U.S. Green Building Council)
MA	Massachusetts
MA EM	Massachusetts Energy Management
MassDEP	Massachusetts Department of Environmental Protection
MGD	Million gallons per day
MOP	Manual of Practices
MTC	Massachusetts Technology Collaborative
MW	Megawatt
MWh	megawatts per hour
MWRDGC	Metropolitan Water Reclamation District of Greater Chicago
NEIWPCC	New England Interstate Water Pollution Control Commission
NH	New Hampshire
NPV	Net present value
NRNC	Nonresidential New Construction
NRR	Nonresidential Retrofit
NYC	New York City
NYPA	New York Power Authority
NYS	New York State
NYSERDA	New York State Energy Research and Development Authority
NYWEA	New York Water Environment Association
O&M	Operation and maintenance
OEP	Office of Energy and Planning
OPH	Office of Public Health
OWSO	Optimization of Wastewater and Solids Operations
PE	Licensed Professional Engineer
PG&E	Pacific Gas and Electric
PON	Project Opportunity Notice
QuEST	Quantum Energy Services & Technologies, Inc.
R&D	research and development
RBC	Rotating Biological Contactors
RFP	request for proposal
RI	Rhode Island
SAIC	Science Applications International Corporation
SAVE	Society of American Value Engineers
SAWS	San Antonio Water System
SBC	system benefits charge
SCADA	supervisory control and data acquisition

SCDHEC	South Carolina Department of Health and Environmental Control
SCE	Southern California Edison
SCS	Selector Contact Stabilization
SDG&E	San Diego Gas and Electric
SMUD	Sacramento Municipal Utility District
SoCalGas	Southern California Gas Company
SRF	State Revolving Fund
SRLF	State Revolving Loan Fund
SRWTF	Sheboygan Regional Wastewater Treatment Facility
TA	Technical Assistance
TCEQ	Texas Commission on Environmental Quality
Ten States or 10 States Standards	Wastewater treatment facility design standards prepared by the Wastewater Committee of the Great Lakes – Upper Mississippi River Board of State and Provincial Public Health and Environmental Managers
TR-16	TR-16: Guides for the Design of Wastewater Treatment Works, published by the New England Interstate Water Pollution Control Commission
U.S. EPA	United States Environmental Protection Agency
UV	Ultraviolet light
VFD	variable frequency drive
WDNR	Wisconsin Department of Natural Resources
WEF	Water Environment Federation
WERF	Water Environment Research Foundation
WEFTEC	Water Environment Federation Technical Exhibition and Conference
WPCRLF	Water Pollution Control (Clean Water) Revolving Loan Fund
WPEI	Wastewater Process Efficiency Initiative
WQCD	Water Quality Control Division
WQD	Water Quality Division
WTS	Water Technical Sheet
WWPEI	Wastewater Plant Efficiency Improvement Initiative Program
WWTF	Wastewater Treatment Facility
WWTP	Wastewater Treatment Plant

Executive Summary

Approximately 30-35% of a wastewater treatment facility's (WWTF) total cost to provide wastewater service is for energy use. Industry-wide, the water and wastewater industries consume about 3% of the electric energy produced in the United States.[1] Clearly, the wastewater sector and other wastewater treatment providers, such as industry and commercial establishments, need cost-effective solutions for energy management and efficiency in their wastewater treatment operations.

The interrelated problems of rising energy costs and demand alongside climate change have brought the issue of energy efficiency to the forefront for most wastewater treatment providers today. The Operations Optimization Challenge is the Water Environment Research Foundation's (WERF's) program to provide integrated, comprehensive, long-term research that addresses process optimization, energy efficiency, energy and resource recovery, and minimization of the environmental footprint of wastewater collection and treatment facilities. The primary objective of the Optimization Challenge is to develop approaches that allow the wastewater sector to attain a 20% or greater reduction in energy without compromising treatment goals.

This report evaluates the feasibility of establishing a national design standard for WWTFs that incorporates energy efficiency-related concepts and provides suggested language for incorporating energy efficiency concepts into design guidelines or standards. Suggested language includes a requirement for scalable system design which can be operated incrementally at flows below maximum design. The suggested language also calls for life-cycle cost analysis (LCCA) for each alternative and to present the net present value of any energy savings.

The report includes results from a technical working session of selected municipal wastewater agencies that focused on the issue of incorporating energy guidelines into their respective standards.

One goal of the Operations Optimization Challenge is to identify the program management best practices that the most effective power utilities or state energy efficiency programs use to address wastewater facilities. This study highlights three states (California, New York, and Wisconsin) that have effective energy reduction programs for the wastewater sector with a long-term history of performance. The exemplary programs showed substantial commonality in best practices. Key features of these programs include the following:

- Implementing energy efficiency measures using a performance-based contract with the program implementation contractor.
- Maintaining lists of qualified engineering consultants and project implementation contractors.
- Supporting a variety of marketing and outreach activities.

---

[1] Carns, Keith, "Bringing Energy Efficiency to the Water and Wastewater Industry: How Do We Get There?", *Proceedings of the Water Environment Federation*, Water Environment Federation Technical Exhibition and Conference (WEFTEC) **2005**: *Session 91 through Session 100*, pp. 7650-7659.

- Providing personalized project facilitation services throughout the project to ensure that it moves to completion and considering providing more program assistance to smaller wastewater systems, which have fewer financial and staff resources than larger wastewater systems have to commit to projects.
- Providing access to independent funding sources and financial incentives such as cost-share grants to develop energy efficiency projects.
- Presenting non-energy benefits of energy efficiency projects (e.g., reduced green house gas emissions and improved sustainability) to municipal decision makers.
- Continuing to respond to participants' needs as the program grows by refining program services.

CHAPTER 1.0

INTRODUCTION

Approximately 30-35% of a WWTF's total cost to provide wastewater service is for energy use. Industry-wide, the water and wastewater industries consume about 3% of the electric energy produced in the United States.[2] Clearly, the wastewater sector and other wastewater treatment providers, such as industry and commercial establishments, need cost-effective solutions for energy management and efficiency in their wastewater treatment operations.

1.1 Background

The interrelated problems of rising energy costs and demand alongside climate change have brought the issue of energy efficiency to the forefront for most wastewater treatment providers today. The Operations Optimization Challenge is the WERF's program to build integrated, comprehensive, long-term research into process optimization, energy efficiency, energy and resource recovery, and minimization of the environmental footprint of wastewater collection and treatment facilities.

1.2 Overview

One goal of the Operations Optimization Challenge is to identify the most effective approaches for incorporating energy efficiency and energy management best practices within the wastewater treatment sector. To this end, the study documented in this report focuses on two targeted activities: 1) developing energy efficiency and energy management guidelines for potential adoption by state organizations charged with setting standards and guidelines for the design of WWTFs and 2) developing a program management framework for establishing a successful utility or state energy efficiency program targeting WWTFs. This framework is based on the results of existing state or utility programs that have been identified as exemplary. This study highlights three states (California, New York, and Wisconsin) that have effective energy reduction programs for the wastewater sector with a long-term history of performance. The objective of this report is to increase the number of states or other agencies (interstate agencies, as well as power providers) that can implement successful programs to help the wastewater sector reduce nationwide energy demand.

1.2.1 Energy Efficiency and Wastewater Facility Design

States, or sometimes interstate groups, publish design standards and recommend guidelines that apply to municipal wastewater treatment plants (WWTPs) during the planning and design phases of facility construction. While some states have energy programs that are well synchronized with their planning and design guidelines, many do not. In addition, many design guidelines suggest that facilities should be planned and designed to meet maximum loads over a 20-year period, which may result in overcapacity and inherent energy inefficiency in the early years of operation.

---

[2] Carns, Keith, "Bringing Energy Efficiency to the Water and Wastewater Industry: How Do We Get There?", *Proceedings of the Water Environment Federation*, WEFTEC **2005**: *Session 91 through Session 100*, pp. 7650-7659.

Chapter 2.0 of this report discusses the status of state design guidelines and their ability to address energy management and energy efficiency requirements. It also provides text for a draft model standard and guideline for ensuring that energy efficiency is considered in WWTF design. Furthermore, the report provides the results of a review of this draft standard and guideline and other considerations by selected wastewater treatment organizations.

1.2.2 Energy Demand and Demand Reduction for the Wastewater Sector

A broad spectrum of energy reduction programs is in place and functioning in the United States. At the state level, several states (New York, Wisconsin, and California) have premier programs that align with United States Environmental Protection Agency (U.S. EPA) energy efficiency and recovery objectives, while operating at the grass-roots level in their respective states. However, no resources exist that categorize or define these programs and their ability to promote, improve, and reduce wastewater-related energy consumption. This report includes an analysis of best practices in these state and power utility programs, as well as a database of these resources. The objective of Chapter 3.0 is to identify the most successful features of state and power utility wastewater sector energy reduction program elements and to provide a framework or template for a model wastewater sector energy reduction program that other agencies can implement.

CHAPTER 2.0

DESIGN GUIDELINES AND STANDARDS

2.1 Objectives

The objectives of this chapter are to provide information on agency status with regard to existing design guidelines and standards that address energy and to provide draft model design guideline and standard wording that addresses energy management practice requirements and energy efficiency.

2.2 Background and Approach

Many states, and in some cases groups of states (e.g., Ten States Standards), have design standards and recommended guidelines that apply to WWTFs. Engineers use these standards during facilities planning and design. Each state's environmental regulatory authorities refer to the standards when reviewing planning-level engineering reports (facilities plans) and construction documents. Several states have energy conservation, energy reduction, and renewable energy programs. In some cases, these states synchronize their energy conservation programs with WWTF design guideline and standard documents. In many cases, the design guidelines and standards do not address energy efficiency or production and use of renewable energy. Because many design guidelines suggest that facilities should be planned for a 20-year life, a benefit of a design guideline that addresses energy efficiency would be to encourage project developers to estimate annual long-term energy savings up front.

This study included the following activities:

1. Contacted state and federal agencies with responsibility for developing and implementing WWTF design guidelines. Determined, either through direct response from the agency or from review of the design guideline, whether the guideline included energy management practices or energy efficiency requirements. Contacted agencies using a combination of e-mail and telephone calls.
2. Compiled design guideline and standard information into a Microsoft Excel spreadsheet database.
3. Developed a draft model guideline and standard that incorporated energy management practices and energy efficiency requirements, modeled after design guidelines with energy-related requirements identified in Activity 1.
4. Distributed the draft model guideline and standard to a select group of states and organizations for review and comment. Discussed the draft guideline with these organizations and compiled their comments.
5. Developed an outline for an Energy Best Practice Guidebook.

2.3 Database of State and Federal Government Design Guidelines

The database of state and federal design guidelines and standards for WWTFs (Appendix A) was compiled in Microsoft Excel (filename: WERF Subtask 1-2 50 States WWTF Design

Guidelines.xls). Some states in the database are represented with more than one line entry. Such cases occurred when the study team made more than one contact with a state to discuss different aspects of WWTF design review. For example, in Alabama, discussions of design guideline perspectives involved 1) the permitting office of the Water Division municipal branch within the state's Department of Environmental Management (DEM) and 2) the Clean Water State Revolving Fund (CWSRF) Loan Program within the DEM. In California, the study team obtained information from the State Water Resources Control Board's Division of Water Quality and CWSRF Program and from the California Energy Commission.

Table 2-1 defines the fields in this database.

Table 2-1. 50 States WWTF Design Guidelines Database Field Definitions.

Field	Definition
State	Self-explanatory
Agency	Name of state agency with responsibility for WWTF design guidelines and permitting
Name	Name and managerial role of agency contact
Telephone Number	Telephone number for agency contact
E-mail	E-mail address for agency contact
Mailing Address	Mailing address for agency
Web Address	Web address for agency
Regulatory Role	For example, WWTF design review, WWTF permitting, or CWSRF
Design Guideline Citation	Indicates whether the state has any design guideline; if it does have a guideline, provides citation and, when available, a website
Energy Management Requirement Content in Current Design Guideline	Yes or no; if yes, describes the content
Exemplary or Best Practice Energy Management Content in Current Design Guideline	Comments on whether an associated state energy program (e.g., New York or Wisconsin) contains such content, even though the content may not be reflected in current WWTF design guidelines issued by the regulatory agency

2.4 Energy Efficiency and Energy Management Practices in State and Federal Design Guidelines

Many states have WWTF design guidelines, but only three (Pennsylvania, Virginia, and Washington) out of 35 respondents reflect energy efficiency or energy management practices in the guidelines, as identified in Table 2-2. The following paragraphs summarize the approaches of Pennsylvania, Virginia, and Washington. Guidelines for Pennsylvania and Virginia address energy efficiency for specific wastewater treatment processes. In contrast, the Washington guideline addresses energy efficiency more generally, stating that the facility plan alternatives analysis should address technologies that save energy. Code changes under consideration in Wisconsin also follow a more general approach of inserting the phrase "energy efficiency" in various code sections.

The energy efficiency language from these three states' design guidelines, provided in the following paragraphs, was considered in drafting a model guideline and standard (Section 2.5).

Pennsylvania – Section 43 (Design):43.1 (Type of Treatment) cites "energy requirements" among "other considerations" for design. Section 82 (Activated Sludge):82.113 (Energy Requirements) recognizes aeration as a major energy demand in the activated sludge process. Additionally, "Energy costs and potential mandatory emergency public power reduction events in relation to critical water quality conditions must be carefully evaluated. Capability of energy usage phasedown while still maintaining process viability, both under normal and emergency energy availability conditions must be included in the activated sludge design". Section 82.332.e (Diffused Air Systems) states "The blowers shall be provided in multiple units, so arranged and in such capacities as to meet the maximum air demand with the largest unit out of service. The design shall also provide for varying the volume of air delivered in proportion to the load demand of the plant. The use of timers to conserve energy is acceptable, provided necessary minimum dissolved oxygen level is maintained in the aeration tank. Aeration equipment shall be easily adjustable in increments and shall maintain solids in suspension with these limits. Step type proportioning is acceptable". Section 83 (Rotating Biological Contactors [RBC]):83.3 (Design Considerations):83.31 (Process Loading Reliability): 83.31.n states "The use of high-efficiency motors for mechanically driven RBC units to lower the energy consumption should be considered" (*Domestic Wastewater Facilities Manual*, Document ID 362-0300-001, October 1, 1997).

Virginia – 9VAC25-790-550 (Anaerobic digestion) F. Energy Control states "If practical, digesters should be constructed above the water table and should be suitably insulated to minimize heat loss. The use of digester gas as a heating fuel source is encouraged". 9VAC25-790-770 (Ultraviolet light [UV] irradiation) states "E. Dose control. For treatment works with a design average daily flow of one mgd or more, UV system design should include a control system to turn appropriate lamps on or off in order to conserve energy" (*Sewage Collection and Treatment Regulations* [9VAC25-790], authorized under the Code of Virginia 1950, as amended, Title 62.1, Section 62.1-44.19 [http://www.deq.state.va.us/wastewater/regulations.html]).

Washington – Requirements for Engineering Report and Facility Plan state that the facility plan alternatives analysis is subject to the same requirements as the engineering report "plus description of innovative and alternative technologies [that is, those saving energy and nonconventional treatment (land application, etc.)]" (*Criteria for Sewage Works Design (Orange Book)*, http://www.ecy.wa.gov/biblio/9837.html, January 1978 [revised August 2008], Table G1-1, p. G1-18).

In some regions and states, authorities are making initial efforts to consider adding energy efficiency language to design guidelines. For example, current efforts are underway to update TR-16, the code that governs WWTF design for New England Interstate Water Pollution Control Commission (NEIWPCC) member states, which include Connecticut, Maine, Massachusetts, New Hampshire, New York, Rhode Island, and Vermont. In September 2008, the Executive Committee of the NEIWPCC voted to update the TR-16 wastewater system design guideline to incorporate energy efficiency. The update of the current 1998 version of TR-16 will take 1-1.5 years. In Wisconsin, authorities are considering the addition of the phrase "energy efficiency" for Wisconsin Administrative Code sections NR809, NR810, and NR811, which address clean water. As of October 2009, this effort is at the public hearing stage. The State of Minnesota Sustainable Building Guidelines, Version 2.0, Energy & Atmosphere Section (http://www.msbg.umn.edu/energy.html), while not addressing WWTFs, is an effort to codify energy efficiency practices. In the future, these cited efforts may lead to additional models for design guideline language.

Table 2-2. Energy Efficiency Language in Design Guideline.

State	Has Design Guideline	Energy Efficiency Language in Design Guideline	State	Has Design Guideline	Energy Efficiency Language in Design Guideline
Alabama	N	N	Maryland	NR	NR
Alaska	Y	N	Massachusetts	Y (TR-16)	N
Arizona	NR	NR	Michigan	Y (Ten States)	N
Arkansas	Y (Ten States)	N	Minnesota	Y (Ten States)	N
California	N	N	Mississippi	NR	NR
Colorado	NR	NR	Missouri	NR	NR
Connecticut	Y (TR-16)	N	Montana	Y	N
Delaware	NR	NR	Nebraska	Y	N
District of Columbia	NR	NR	Nevada	Y	N
Florida	Y	N	New Hampshire	Y	N
Georgia	Y	N	New Jersey	Y	N
Hawaii	NR	NR	New Mexico	Y	N
Idaho	Y	N	New York	Y (Ten States)	N
Illinois	Y		North Carolina	NR	NR
Indiana	Y (Ten States)	N	North Dakota	NR	NR
Iowa	Y (Ten States)	N	Ohio	Y (Ten States)	N
Kansas	N	N	Oklahoma	NR	NR
Kentucky	NR	NR	Oregon	N	N
Louisiana	NR	NR	Pennsylvania	Y	Y
Maine	Y (TR-16)	N	Rhode Island	Y (TR-16)	N
South Carolina	Y	N	Virginia	Y	Y
South Dakota	NR	NR	Washington	Y	Y
Tennessee	NR	NR	West Virginia	NR	NR
Texas	Y	N	Wisconsin	Y Chapter NR 110 Wisconsin Administrative Code (supplementary use of Ten States)	N
Utah	Y	N	Wyoming	Y	N
Vermont	Y (TR-16, Ten States)	N	Federal–U.S. EPA	NR	NR

Notes:

Y: Yes N: No NR: No survey response

TR-16 – New England Interstate Water Pollution Control Commission, *TR-16 Guides for the Design of Wastewater Treatment Works*, http://www.neiwpcc.org/tr16guides.asp, 1998.

Ten States – Wastewater Committee of the Great Lakes – Upper Mississippi River Board of State and Provincial Public Health and Environmental Managers, *Recommended Standards for Wastewater Facilities*, http://10statesstandards.com/wastewaterstandards.html, 2004. Member states include Illinois, Indiana, Iowa, Michigan, Minnesota, Missouri, New York, Ohio, Pennsylvania, and Wisconsin.

2.5 Draft Model Guideline and Standard

In Section 2.4, code examples for Pennsylvania and Virginia show a technology-specific approach to addressing energy efficiency in wastewater treatment subsystems. The Washington code presents a more general approach to addressing energy efficiency that applies to the whole wastewater system. In Wisconsin, authorities are considering a similar approach emphasizing systemwide energy efficiency alternatives. The more general Washington approach served as a starting point for developing a draft model guideline and standard because the intent of the wording is to encourage design engineers to consider energy-efficient alternatives at the wastewater system design level. The draft wording that follows augments the general approach used in Washington to include the concepts of designing a wastewater system to be scalable as system throughput increases over the facility lifetime and to operate in an energy-efficient manner in scaled-up operating modes. Additionally, the wording augments the Washington approach by requiring life-cycle cost analyses (LCCA) to evaluate energy efficiency benefits. This level of analytical detail is consistent with design guideline requirements included in Public Law 92-500, Federal Water Pollution Control Act (Clean Water Act), which is the basis for many states' wastewater system design guidelines.

Following is the suggested energy efficiency and energy management language for incorporation into WWTF design guidelines and standards:

> *All projects requiring a facility plan, facility plan amendment, comparison of alternatives, facility modification, or a facility re-evaluation for capacity rating shall:*
>
> *1) include system design alternatives that are incrementally scalable as system throughput increases over the facility lifetime. To be incrementally scalable, system designs shall consider installation of multiple units of a system component (e.g., drive, motor, pump, aeration facility blower) such that (a) when operated together, all component units can be operated in an energy-efficient manner at the maximum design loading specifications and (b) when operated incrementally at loadings below the design specification, the component units can also be operated in a way to maximize energy efficiency.*
>
> *2) identify three project alternatives with high net present value (i.e., considering capital and operating costs) based on life-cycle cost analyses (LCCA) and for each, calculate the net present value of energy savings. The energy savings attributable to each process component (e.g., drive, motor, and method of control) shall be evaluated separately and together as a whole system (all components integrated). The LCCA shall evaluate the energy savings attributable to incrementally scalable system design alternatives defined in (1), above, and shall include a sensitivity analysis that considers how energy supply cost volatility may affect the LCCA.*

2.6 Agency Comments on Draft Model Guideline and Standard

The draft model guideline and standard presented in Section 2.5 was circulated to selected organizations (identified in Table 2-3) that currently do not have energy efficiency language in their design guidelines. The selections emphasize municipal agency contacts, rather than state- or regional-level contacts, because experience in Wisconsin has shown the grass roots approach to be most effective in building statewide support for energy efficiency language in design guidelines. The selections represent a geographically diverse group. Two selections (Great Lakes – Upper Mississippi River Board of State and Provincial Public Health and Environmental Managers and NEIWPCC) represent WWTF standards organizations with multistate coverage. Contact with NEIWPCC was deemed useful because they are working on energy efficiency wording as part of an update of their TR-16 design guidelines. The study included agencies that currently benefit from a statewide WWTF energy efficiency program (e.g., Wisconsin and New York). The market transformation hypothesis is that state programs already emphasizing WWTF energy efficiency may be the most receptive to incorporating energy efficiency language into design guidelines.

Table 2-3. Organizations for Consideration of Model Energy Efficiency Language in WWTF Design Guidelines/Standards.

State/Organization	Exemplary WWTF Energy Efficiency Program	Current Design Guideline Citation
1. New York	NYSERDA Focus on Water & Wastewater Program	Ten States Standards
2. Wisconsin/Green Bay Metropolitan Sewage District	Focus on Energy Water & Wastewater Program	Chapter NR 110 Wisconsin Administrative Code
3. Wisconsin/Sheboygan	Focus on Energy Water & Wastewater Program	Chapter NR 110 Wisconsin Administrative Code
4. Great Lakes – Upper Mississippi River Board of State and Provincial Public Health and Environmental Managers	None identified	Ten States Standards
5. NEIWPCC	Training sessions on reducing energy usage at WWTFs for cost savings.[1] Current effort to update TR-16 to include energy efficiency wording. NEIWPCC member states such as Connecticut, Maine, Massachusetts, and New York have programs to fund WWTF energy efficiency projects	TR-16 Guides for the Design of Wastewater Treatment Works
6. Washington/Seattle – King County	None identified	State of Washington, Department of Ecology, *Criteria for Sewage Works Design (Orange Book)*, http://www.ecy.wa.gov/biblio/9837.html, January 1978 (revised August 2008)
7. Florida/Miami – Dade County	None identified	Rule 62-600 Florida Administrative Code (F.A.C.)/Domestic Wastewater Facilities, Chapter 62-600.400 (Design Requirements), http://www.dep.state. fl.us/legal/Rules/wastewater/62-600.pdf; F.A.C., Guide to Permitting Wastewater Facilities or Activities under Chapter 62-620, http://www.dep.state.fl.us/water/wastewater/dom/docs/wwguide.pdf
8. Illinois/Metropolitan Water Reclamation District of Greater Chicago	Electric utilities such as Exelon and Ameren fund energy efficiency projects for WWTFs	Illinois Environmental Protection Agency, Bureau of Water, http://www.epa.state.il.us/water/rules-regulation.html
9. Texas/Dallas	None identified	Texas Commission on Environmental Quality, *Wastewater System Design Requirements*, http://www.tceq.state.tx.us/permitting/water_quality/wastewater/plans/WW_design_requirements.html, 2008
10. Texas/San Antonio	None identified	Texas Commission on Environmental Quality, *Wastewater System Design Requirements*, http://www.tceq.state.tx.us/permitting/water_quality/wastewater/plans/WW_design_requirements.html, 2008

Notes:

[1] New England Interstate Water Pollution Control Commission, *2007 Annual Report*, http://www.neiwpcc.org/neiwpcc_docs/AR2007.pdf, p. 6.

Wastewater treatment organizations were invited to provide written comments on the draft model guideline and standard wording and were invited to participate in a teleconference to discuss the draft wording. Comments fall into three categories:

1. When should an evaluation of energy efficiency alternatives be required? – Roger Browne (King County) suggests that such evaluations should not apply to "a facility re-evaluation for capacity rating".
2. System performance wording – Roger Browne (King County) and Dale Doerr (Sheboygan) stated wording preferences for "loading" and "throughput", which appear in the draft wording. Participants discussed whether an "incrementally scalable" WWTF could be designed such that "(2) when operated incrementally at loadings below the design specification, the component units can **also** be operated in a way to maximize energy efficiency". Dan Busch (Green Bay) emphasized in his written comments that an incrementally scalable design must address system reliability, redundancy, and maximum energy efficiency, collectively.
3. Economic burden of developing multiple design options – Chris Nam (Chicago) indicated that up-front design costs limit the number of evaluations and that a requirement to evaluate three project alternatives would present an economic burden.

Based on the written comments and discussion during the teleconference, the following presents alternative draft model guideline and standard wording:

> ***Alternative Draft Wording:*** *All projects requiring a facility plan, facility plan amendment, comparison of alternatives, facility modification, or a facility re-evaluation for capacity rating shall:*
>
> *1) include system design alternatives that are scalable as system throughput increases such that when operated together, component units and major subsystems can be operated in an energy-efficient manner over the expected facility lifetime. To be scalable, system designs shall consider installation of multiple units of a system component (e.g., drive, motor, pump, aeration facility blower).*
>
> *2) all project alternatives shall be subjected to a life-cycle cost analysis (LCCA) that considers capital and operating cost in addition to functional requirements, such as process efficiency, reliability, and permit compliance. The energy usage attributable to each process component (e.g., drive, motor, and method of control) shall be evaluated separately and together as a system with all of its components integrated. The LCCA shall evaluate the energy usage attributable to scaled system design alternatives defined in (1) above and shall include a sensitivity analysis that considers how energy supply cost volatility may affect the LCCA. Energy to operate treatment works is a precious resource. Where other project considerations are equal, give preference to lower energy usage alternatives and projects that recover energy resources.*

Organizations participating in the teleconference also discussed the following two topics:

1. Comments regarding the need for design guideline energy efficiency wording. Would such wording support regional, state, and local initiatives to advance energy efficiency?

Currently, participants are implementing energy efficiency options as a standard practice in wastewater treatment system design. For King County, energy efficiency wording is in the county's Orange Book, which governs WWTF design. In Green Bay, Chicago, and San Antonio, energy efficiency considerations are standard practice without the force of energy efficiency wording in design guideline code. Energy efficiency project implementation may be impeded by factors that include 1) oversizing capacity relative to near-term loads so that the wastewater system can handle higher loads over a 20-plus-year system life, 2) building redundancy into a wastewater system to ensure reliability, and 3) implementing low bid rules that make it more difficult to implement more expensive energy efficiency alternatives.

2. Pathways to incorporating energy efficiency wording into regional, state, and local design guidelines, including barriers to incorporating wording and approaches to addressing barriers.

Participants discussed the State Revolving Fund (SRF) as a pathway to implementing energy efficiency projects. Specifically, energy efficiency project implementation could be assisted by 1) including energy efficiency as an SRF loan evaluation criterion, 2) offering lower SRF loan interest rates for energy efficiency projects, and 3) lowering the principal to be repaid on an energy efficiency project loan through principal forgiveness authorized by the SRF. The SRF pathway is a possible adjunct to the design guideline energy efficiency wording approach.

Appendix B presents the agency meeting report for this teleconference, including the list of attendees. Seven of the 10 organizations listed in Table 2-3 attended the teleconference. Post-teleconference follow-up included distributing the agency meeting report to all organizations listed in Table 2-3 and inviting final comments regarding the draft model standard and guideline wording and related discussion topics.

The organizations were asked to address five questions regarding how they might be affected by the addition of energy efficiency wording to design guidelines and what pathways and barriers they perceived regarding the addition of energy efficiency wording to design guidelines. Their responses to the following five questions provide insight for consideration:

1. How might such wording affect your organization's wastewater treatment facility design effort?

♦ Some of the organizations indicated that consideration of energy-efficient design alternatives already is standard practice.

♦ Requiring a facility plan to include incrementally scalable system design alternatives and to calculate the net present value (NPV) of energy savings for high NPV project alternatives would eliminate internal debate within the wastewater treatment organization regarding inclusion of energy efficiency considerations in design alternatives.

♦ Smaller wastewater treatment operations may not be addressing energy efficiency; such wording would affect their design efforts.

2. How might such wording affect management of your organization's wastewater treatment facilities?

♦ Some of the organizations already consider energy-efficient design alternatives as part of their standard practice; they indicated that the addition of energy efficiency wording to design guidelines would not affect facilities management.

♦ Management would be required to include energy efficiency as a design criterion. Without such a requirement, reliability, redundancy, designing to meet maximum 20-year loads, and cost considerations can lead to less energy-efficient designs.

3. Would such wording support regional, state, and local initiatives to advance energy efficiency?

♦ Answers to this question addressed the other four questions.

4. What pathways might be used to incorporate energy efficiency language into regional, state, and local design guidelines?

♦ Consider making energy efficiency a requirement for facility plan approval.

♦ An alternative to adding energy efficiency wording into design guidelines would be for the state revolving loan fund (SRLF) to reward energy efficiency projects with reduced loan rates, providing municipalities with a cost-savings incentive for pursuing energy efficiency. Such savings is a tangible monthly benefit; each month, the municipality makes a lower loan payment.

♦ In Washington, the pathway would start at the Washington State Department of Ecology. This department issues policy and guidance documents. State-level policy statements are needed to express state-level interest in developing energy-efficient design guidance.

5. What issues or barriers need to be addressed before incorporating energy efficiency wording into design guidelines, and what are your suggested approaches to addressing these barriers?

♦ Payback on energy efficiency projects may exceed a municipality's acceptance guidelines.

♦ Design engineers need to become more comfortable with incorporating energy efficiency and incremental systems into designs. Currently, designs are conservative; they are affected by considerations for reliability and safety that result in less energy efficiency and more redundancy.

♦ Many facilities do not address energy efficiency alternatives in their design efforts because of barriers such as resource constraints, limited staff time, and limited expertise in energy-efficient alternatives. A possible pathway to addressing these barriers would be to emulate Energy Star branding, reinterpreted for wastewater treatment systems. A professional industry association could develop wastewater treatment equipment or system ratings that emulated the Energy Star system. Then, WWTF designers would be able to consider an endorsed system as a design alternative.

2.7 Water and Wastewater Industry Energy Best Practice Guidebook

An Energy Best Practice Guidebook, identifying energy-efficient best-practice options for wastewater treatment system design, provides a resource for implementing the energy-efficient design requirements specified in draft model guideline and standard language. The guidebook developed for the Wisconsin Focus on Energy Program is presented as an example of a best-practice resource. The Wisconsin Focus on Energy Program has given WERF permission to make this guidebook available to its members. Appendix C provides an outline of relevant sections in the Wisconsin Focus on Energy Best Practice Guidebook and provides web links to the guidebook. This guidebook will help municipalities implement sound energy management practices regardless of whether guideline wording is in place.

CHAPTER 3.0

STATE ENERGY REDUCTION PROGRAM RANKING AND BEST PRACTICES

3.1 Objectives

This chapter provides a framework for a model wastewater system energy reduction program that state agencies or utilities can implement.

3.2 Background and Approach

A broad spectrum of energy reduction programs is currently operational in the United States. At the federal level, the U.S. EPA Energy Star program identifies and promotes energy-efficient products, facilitates technology transfer, and promotes energy reduction through thousands of partnering agreements. At the state level, California, New York, and Wisconsin are examples of premier programs that align with the U.S. EPA program, while also operating at the grass-roots level. At the utility level, a large number of regional utilities manage energy reduction programs and offer incentives to rate payers for implementing best energy management practices. Such programs may have a direct or indirect impact on energy management–related activities implemented within wastewater systems. Several wastewater treatment systems have exemplary programs as a result of state- and utility-based energy reduction initiatives. Chapter 3.0 provides a framework for a model program for wastewater system energy reduction based on the best practices for exemplary state- and utility-based wastewater system energy reduction programs.

The program framework followed a four-step research approach:

1. Defined categories to organize wastewater system energy reduction program information for comparing key program features. The following categories characterize programs: year of program inception; number of wastewater systems that have benefited from the program; number of energy reduction projects initiated or completed under the program; and total annual electrical energy savings, all projects (Table 3-1, Section 3.3). Additionally, implementing legislation, agency structure, program management, technical outreach, education and training, and research and development (R&D) projects were factors of consideration (data, as available, presented in Appendix D).
2. Obtained and summarized state utility wastewater system energy reduction program information according to the categories established in Step 1, and with emphasis on program best practices. Appendix D contains program information.
3. Used information in Step 2 to identify the most successful program features (e.g., technology transfer materials, outreach programs, and strategies that led to wastewater systems' increased use of the program to initiate energy reduction projects).
4. Used program best practices identified in Step 3 to develop a framework for a model wastewater-related statewide program.

3.3 Analysis of State Wastewater Treatment Facility Energy Reduction Programs

An analysis of existing wastewater system energy reduction programs identified program management best practices. The selected programs were from existing multiprogram comparative studies of program management best practices[3] and from a 50-state survey of state-level wastewater system regulatory agencies conducted to identify state-level wastewater system design guidelines (Chapter 2.0). The multiprogram studies analyzed the California Wastewater Process Optimization Program, NYSERDA programs, Wisconsin Focus on Energy Water and Wastewater Program, and the Pacific Gas & Electric (PG&E) program. Table 3-1 provides performance metrics for some programs, highlighting the number of projects and aggregate energy savings.

The programs analyzed in the study share many of the same program management best practices, suggesting that, when a new program starts, its managers learn from the experience of existing programs. Because of this learning curve, analysis of additional programs may not yield significantly different best practices. Indeed, the Bonneville Power Administration (BPA) comparison of programs resulted in a synthesis of program management best practices that can serve as a basis for its wastewater program design.

[3] The two studies are John Bidwell (QuEST), Jennifer Fagan (Itron), Todd Amundson (BPA), "Water-Wastewater Program Best Practices: Lessons from the Field", presented at American Council for an Energy-Efficient Economy (ACEEE) 2008 Summer Study on Energy Efficiency in Buildings, Pacific Grove, CA, August 2008, which compares the California Wastewater Process Optimization Program (CalPOP) and BPA programs, and the associated BPA study, which analyzed program management best practices based on the NYSERDA, PG&E, and Wisconsin Focus on Energy Water and Wastewater programs.

Table 3-1. Various Wastewater Treatment Facility Energy Reduction Program Metrics.

Sponsoring Organization	Program	Year of Program Inception	Number of Wastewater Systems Using Program	Number of Projects	Electrical Energy or Demand Savings
California Public Utility Commission	California Wastewater Process Optimization Program	2000[1]	More than 50 (2000–2005)[1]	34 (2002–2008) [1]	More than 25 GWh (2000–2005)[1] Goal 4.5 GWh (2006–2008)[1]
Massachusetts Department of Environmental Protection	Energy Management Pilot for Wastewater and Drinking Water Plants	2007[2]	7[2]	7[2]	14.9 GWh per year[3]
New York Power Authority	Anaerobic Digester Gas-Fueled Fuel Cells Projects	*	5	5	1.8 MW total capacity
New York State Energy Research and Development Authority	Focus on Water and Wastewater Program	2007	*	Number of Projects Approved through September 30, 2008:[4] Energy Efficiency Improvements at Water and Wastewater Treatment Plants: 1 Submetering: 2 Demonstrations: 16 Technical Assistance: 81 Technology Transfer: 1	*
Wisconsin Focus on Energy	Water and Wastewater Program	2000[5]	More than 200 municipal and private facilities	*	Greater than 140 GWh potential savings identified (1999–2003)[5]

Notes:

*No response or no data available.

1. John Bidwell (Quantum Energy Services & Technologies, Inc. [QuEST]), Jennifer Fagan (Itron), Todd Amundson (BPA), "Water-Wastewater Program Best Practices: Lessons from the Field", presented at ACEEE 2008 Summer Study on Energy Efficiency in Buildings, Pacific Grove, CA, August 2008, p. 6. Project data are not available for the initial project years (2000–2001). For the 2006-2008 program cycle, eight projects were initiated and completed (e-mail communication to William R. King (Science Applications International Corporation [SAIC]) from Barbara Dilts (PG&E), November 2009).
2. Massachusetts Department of Environmental Protection, Massachusetts Energy Management Pilot, http://www.mass.gov/dep/water/wastewater/empilot.htm, September 2009.
3. Michael DiBara, Massachusetts Energy Management Pilot for Water & Wastewater Plants, EPA New England National Sustainable Infrastructure Meeting, November 6, 2008, MA EM Pilot EPA Sustainability_11-06-08_2 (shrunk).ppt, slide 18.
4. New York Energy $martsm Program, Quarterly Evaluation and Status Report, quarter ending September 30, 2008, Final Report, http://www.nyserda.org/publications/3rd%20Quarter%202008%20Report.pdf, November 2008. Section 5.11 (Municipal Water and Wastewater Efficiency), Table 5-18.
5. Kristi Kezar (SAIC), Joseph Cantwell (SAIC), "Lessons in Designing and Delivering Energy Efficiency Programs for the Water and Wastewater Industry", white paper, p. 2.

In addition to the basic program information obtained by contacting each of these organizations, the study team relied significantly on the results of a study commissioned by the BPA to analyze wastewater program management best practices. BPA commissioned the best practices study as part of its planning effort to implement a wastewater energy efficiency program. The analysis included interviews with managers of the following water and wastewater programs, in addition to web-based information on four other programs:[4]

- CalPOP
- Wastewater Plant Efficiency Improvement Initiative Program (WWPEI) – implemented by KEMA in the PG&E service territory.
- Wisconsin Focus on Energy Water and Wastewater Program
- NYSERDA – Wastewater Initiative.

Review of these programs yielded a set of program best practices for program design, set-up, and implementation (Table 3-2). The review shows that several programs already share many best practices. For example, the review shows that best practices identified for CalPOP are shared by the broader set of programs included in the BPA analysis. The Wisconsin Focus on Energy Water and Wastewater Program best practices (Table 3-3) share similarities with those recognized by the BPA analysis, and they can be analyzed in terms of the same program functions (e.g., project identification and lead generation, project screening, and project facilitation). Within these program function categories, the Wisconsin program expands best practices to address the following examples of best practices in the framework that extend beyond those covered by Bidwell, Fagan, and Amundson (i.e., Table 3-2) include the following:

- **Project Screening** Favoring incrementally scalable system designs and distributing self-evaluation tools to participants
- **Project Facilitation** Providing more program assistance to smaller, rather than larger, wastewater systems. In-house technical personnel working for larger systems (10 million gallons per day [MGD] or larger) may lack knowledge of innovative energy efficiency practices and technologies, but large systems have the financial resources to pay consulting firms with such knowledge to identify opportunities and conduct preliminary assessments. After identification of measures, engineering staff for such systems are large enough to support planning and design; or such systems have working relationships with consulting firms that can assist. Small and medium systems (less than 10 MGD), especially those in rural areas, have limited engineering staff, with little or no in-house technical expertise to identify energy reduction measures. Consulting firm services are not as affordable to small and medium systems because of budget constraints. Thus, free or subsidized project identification, planning, and design services offered by an energy reduction program enable small and medium systems to pursue energy efficiency projects.
- **Project Funding and Viability** Using financial incentives, such as feasibility study grants to analyze energy-efficient alternatives, to get end users to consider energy-efficient technologies. Grants also indicate program support for a project. Such support may be the

---

[4] John Bidwell (QuEST), Jennifer Fagan (Itron), Todd Amundson (BPA), "Water-Wastewater Program Best Practices: Lessons from the Field", presented at ACEEE 2008 Summer Study on Energy Efficiency in Buildings, Pacific Grove, CA, August 2008.

deciding factor in local decisions to invest, boosting management confidence that it is making the right decision. Streamlining the funding process, with attention to financial assistance requirements and paperwork reduction.

- **Project Life Cycle Cost Management** This category is useful based on Wisconsin program experience because it recognizes that program actions are necessary throughout the program's life to ensure against future reversal of current energy-efficient practices and that the program continues to assess participants' needs and new technology opportunities and to respond with program refinements.

Table 3-2. Best Practices Summary Resulting from BPA Analysis.

PROGRAM DESIGN AND SET-UP	
Implementation Approach	• Use a performance-based contract with the implementation contractor. This provides Implementer with a strong incentive to pursue projects with viable candidates.
Program Management	• If possible, have a dedicated program manager that is knowledgeable about wastewater operations and well-connected. • Provide a wastewater expert that serves in an independent oversight role.
Contractor Selection	• Provide the flexibility to allow customers to use either their own engineering contractor or one under contract to the program. • Maintain a cadre of well-qualified contractors to provide technical assistance as needed.
PROGRAM IMPLEMENTATION	
Marketing and Outreach	• Use training workshops to educate the market actors and end-users. In addition to building market awareness, this can serve as a lead-generator for the program. • Perform outreach via industry conferences and trade association meetings. • Demonstrate the measure's viability via case studies, preferably those that are local or regional.
Project Identification and Lead Generation	• Rely on the market actors to identify and refer candidates to the program.
Project Screening	• Vet candidates by performing site visits and collecting intelligence from trade allies. • Carefully screen candidates, to identify organizations that are proactive, therefore more likely to implement projects.
Project Facilitation	• Provide frequent follow-up to ensure the project is moving and not "stalled." • If possible, use a personalized approach to follow up with the candidate, ideally with a wastewater expert. • Target efforts initially toward operations staff, and later to high-level municipal and public finance authorities.
Project Funding and Viability	• Provide access to independent funding sources for organizations that do not have sufficient capital to fund projects. • Emphasize non-energy benefits as part of the project justification.

Source: John Bidwell (QuEST), Jennifer Fagan (Itron), Todd Amundson (BPA), "Water-Wastewater Program Best Practices: Lessons from the Field", presented at ACEEE 2008 Summer Study on Energy Efficiency in Buildings, Pacific Grove, CA, August 2008, Table 1.

Table 3-3. Wisconsin Focus on Energy Water and Wastewater Program Best Practices.

Program Activity	Best Practice
Project Identification and Lead Generation, Project Screening, and Project Facilitation	**Respond to specific site needs with relevant services; recognize how facility capacity may affect choice of energy efficiency alternatives.** Once the Wisconsin program began to address specific site needs with relevant services, satisfied participants began to recommend the program to other wastewater system operators. These recommendations have led to increased program growth, including expansion of the program across the state. In particular, when the program's energy advisers use information on facility capacity and process complexity to match energy efficiency measures to existing load conditions, they observe the following: sizing to match load improves cost-effectiveness; sizing to match load increases the likelihood that a project will move forward to completion; and a larger number of facilities, of different capacities, across the entire state, are attracted to the program because projects are tailored to their specific characteristics.
Project Identification and Lead Generation	**Promote participant diversity.** The program has begun to target industrial WWTFs across the state, such as those serving paper mills and food processing plants. Additionally, the program increased efforts to network facility staff and energy-efficient equipment and service providers in the area.
Project Screening	**Design incrementally scalable equipment and system alternatives that can be operated to more closely match lower daily loads early in the facility's 20-year life cycle, rather than following the existing practice of designing for 20-year peak load.** Wisconsin program experience indicates that facilities rarely operate at peak, particularly in the early life of the facility. Substantial energy savings are achievable through projects that install equipment sized to match usually experienced lower loads. **Develop strategies for managing increased burden program growth and participant diversity places on project screening resources.** The program distributes and encourages the use of self-evaluation tools and information. This strategy enables the program to respond to as many facilities as possible in a timely manner. It is a useful strategy to address program growth, which places higher demand on program services. The main purpose of these materials is to provide a faster turnaround time for program participation at a lower cost to the program. The self-evaluation methods allow facility staff to focus on projects that they feel deserve top priority.
Project Facilitation	**Consider providing more program assistance to smaller, rather than larger, wastewater systems.** Larger systems (10 MGD or greater) have a larger engineering staff or regularly work with consulting firms. In-house technical personnel working for larger systems may lack knowledge of innovative energy efficiency practices and technologies, but large systems have the financial resources to pay consulting firms with such knowledge to identify opportunities and conduct preliminary assessments. Staff members in small and medium systems (less than 10 MGD), especially those in rural areas, lack the time resources to identify appropriate energy-saving projects. They also often have little or no in-house technical expertise and typically do not have an affiliation with a particular consulting firm. In response to their needs, the program provides more detailed information on the recommended technology, as well as more explanation of the assessment methods and calculations. Feedback from participants in municipal facilities indicated that the detailed written report helped explain the nonmonetary benefits of suggested measures to managers and committees. In many cases, those making final decisions had no technical experience and based their opinions solely on financial considerations.
Project Funding and Viability	**Set financial incentives to promote investment in energy-efficient technology.** Focus on Energy provides cost-share grants to participants that commit to installing energy-efficient equipment. This program management practice addresses the financial barrier to project development, a barrier common to facilities of all sizes, designs, and locations. The program also has grant money available to offset the cost of feasibility studies to provide more detailed assessments of identified projects. Feasibility studies can provide management with a reassuring second opinion that energy efficiency is a worthwhile investment and can indicate which equipment option will be most beneficial financially. Grants also are a strong indicator of program support for a project. Even a small investment on behalf of the program increased management confidence in a project. In several cases, a grant has been the deciding factor for facilities that were uneasy about investing in energy efficiency. **Streamline the funding process.** The program also updated requirements for financial assistance and implemented a faster application process with minimal paperwork.

Table 3-3. Wisconsin Focus on Energy Water and Wastewater Program Best Practices.

Program Activity	Best Practice
Project Life-Cycle Management	**Influence future decisions.** While encouraging the installation of energy-efficient equipment has a direct impact on program participants' operating costs, the benefits of resource acquisition activities are guaranteed only for the lifetime of the equipment. To ensure that the program has long-lasting effects on energy efficiency choices in the water and wastewater industry, services must influence future decisions. Municipalities must take steps to ensure that the benefits of a cost-effective energy efficiency measure installed today are not reversed in the future by replacing failed equipment with less-efficient replacement equipment. They must provide education and training opportunities to explain the reasons and methods of energy efficiency to operators, enabling facility staff to make informed decisions. They must communicate with regulatory agencies to influence future codes and design requirements, including energy efficiency considerations and renewable energy technologies, such as combined heat and power (CHP) using biogas. **Continue to respond to participants' needs.** As the program grows, municipalities must continue to refine and tailor program offerings to meet participant's needs. They must obtain feedback from participants, trade groups, and others in the water and wastewater industry to tailor offerings. Feedback for the Wisconsin Focus on Energy Water and Wastewater Program has been exceedingly positive. Periodically, municipalities must review water and wastewater industry technology offerings for new energy-efficient technology alternatives and develop alternatives to meet wastewater system operator demand as energy costs continue to escalate and state and local governments struggle to cut costs without compromising services.

Source: Based on lessons cited in Kristi Kezar (SAIC), Joseph Cantwell (SAIC), "Lessons in Designing and Delivering Energy Efficiency Programs for the Water and Wastewater Industry", white paper, 2004, pp. 2-4.

3.4 Framework for Model Statewide Wastewater Energy Efficiency Program

Table 3-4 presents a framework for a model program for statewide wastewater energy efficiency. The framework is a synthesis of best practices presented in previous sections of this chapter. The framework is an extension of the analysis presented by Bidwell, Fagan, and Amundson[5] because their best practice framework represents a multiprogram synthesis. Bidwell, Fagan, and Amundson organized best practices according to program functions used (i.e., program implementation categories such as project screening and project facilitation). Extended program functions include project life-cycle management.

Table 3-4. Best Practices Framework for Model Statewide Wastewater Energy Efficiency Program.

Design and Set-Up	
Implementation Approach	• Use a performance-based contract for program implementation
Program Management	• Hire a dedicated program manager who has wastewater operations expertise and a good understanding of electrical equipment and its operation • Ensure that the manager is well connected to regional wastewater operators • Require the program to maintain lists of qualified engineering consultants and project implementation contractors
Contractor Selection	• Allow the customer to use its own engineering contractor or program-approved contractor
Implementation	
Marketing and Outreach	• Perform outreach through industry conferences and trade associations • Demonstrate the measure's viability through case studies, emphasizing local or regional studies and emerging technologies • Use training workshops to educate market suppliers, designers, and end users, the byproduct of which may be generation of leads for the program
Project Identification and Lead Generation	• Rely on market suppliers, designers, and end users to identify and refer candidates • Attract end users to the program with projects tailored to their specific site characteristics (tailored projects generate referrals from satisfied end users) • Promote participant diversity, looking for new market growth opportunities as program grows; for example, expand services beyond municipal WWTFs to industrial WWTFs (e.g., paper mills and food processing plants)
Project Screening	• Help candidates by performing site visits and collecting intelligence from trade allies • Identify organizations that are proactive, thus more likely to implement projects • Favor incrementally scalable system designs that can match lower daily loads early in a facility's 20-year life cycle (such projects are more likely to be completed) • Distribute self-evaluation tools to participants, allowing program staff to focus on more complex, higher-priority projects as the program grows
Project Facilitation	• Provide follow-up throughout project marketing, planning, and implementation to ensure that the project is moving • Use a personalized approach to follow up with the candidate, assigning a wastewater expert to conduct the follow-up • Target efforts initially toward operations staff; later, toward high-level municipal and public finance authorities (during the decision phase) • Consider providing more program assistance to smaller, rather than larger, wastewater systems

[5] John Bidwell (QuEST), Jennifer Fagan (Itron), Todd Amundson (Bonneville Power Administration), "Water-Wastewater Program Best Practices: Lessons from the Field", presented at ACEEE 2008 Summer Study on Energy Efficiency in Buildings, Pacific Grove, CA, August 2008, Table 1.

Table 3-4. Best Practices Framework for Model Statewide Wastewater Energy Efficiency Program.

Project Funding and Viability	• Provide access to independent funding sources for organizations that do not have sufficient capital to fund projects • Set financial incentives to promote investment in energy-efficient technology (e.g., cost-share grant for a feasibility study to identify the best alternative) • Present nonenergy benefits as part of the project justification • Streamline the funding process, with attention to financial assistance requirements and paperwork reduction
Project Life-Cycle Management	• Influence future decisions to ensure that investments in energy-efficient equipment and operations do not reverse at a future date • Continue to respond to participants' needs as the program grows, refining program services and providing new energy-efficient technology alternatives as facility operator requirements and the energy-efficient technology market evolve

CHAPTER 4.0

CONCLUSIONS AND RECOMMENDATIONS

Conclusions address the research efforts to develop draft design guideline and standard wording, and they identify best practices for managing state and utility WWTF energy efficiency programs. Recommendations identify potential future activities to promote adoption of design guideline and standard wording and best practices for managing energy efficiency programs.

4.1 Conclusions

4.1.1 Design Guidelines and Standards

- Few states have WWTF design guidelines and standards that include energy efficiency and energy management considerations. Of the 35 states that responded to the request for information, only three (i.e., Pennsylvania, Virginia, and Washington) included such considerations.
- Pennsylvania and Virginia have design guidelines that address energy efficiency for specific wastewater treatment processes. In contrast, Washington's design guidelines require broader consideration of energy-efficient alternatives.
- Additional states and regions are considering design code revisions to address energy efficiency. Currently, the New England Interstate Water Pollution Control Commission is revising the *TR-16: Guides for the Design of Wastewater Treatment Works* (TR-16) design code for wastewater treatment. Wisconsin is considering changes in the Wisconsin Administrative Code to address energy efficiency in drinking water facility design. Minnesota has taken steps to codify energy efficiency practices for building design.
- Following is the suggested energy efficiency and energy management language for incorporation into WWTF design guidelines and standards that was circulated to selected organizations for comment. Based on the organizations' written comments and discussion during a teleconference, the study team developed alternative draft model guideline and standard wording (see Section 2.6):

All projects requiring a facility plan, facility plan amendment, comparison of alternatives, facility modification, or a facility reevaluation for capacity rating shall:

1) *include system design alternatives that are incrementally scalable as system throughput increases over the facility lifetime. To be incrementally scalable, system designs shall consider installation of multiple units of a system component (e.g., drive, motor, pump, aeration facility blower) such that: (a) when operated together, all component units can be operated in an energy efficient manner at the maximum design loading specifications and (b) when operated incrementally at loadings below the design specification, the component units can* ***also*** *be operated in a way to maximize energy efficiency.*

2) *identify three project alternatives with high net present value (i.e. considering capital and operating costs) based on life cycle cost analyses (LCCA) and for each calculate the net present value of energy savings. The energy savings attributable to each process component (e.g., drive, motor, and method of control) shall be evaluated separately and together as a whole system (all components integrated). The LCCA shall evaluate the energy savings attributable to incrementally scalable system design alternatives defined in (1), above, and shall include a sensitivity analysis that considers how energy supply cost volatility may affect the LCCA.*

- Based on discussions with five municipal wastewater treatment utilities, the study team found mixed support for incorporating energy efficiency language into design code. For utilities that supported the language, the principal reason was that:
    - An energy efficiency design guideline may be necessary to balance design guidelines that tend to decrease energy efficiency (e.g., system design to meet 20-year peak loads, incorporating redundancy into design to ensure system reliability).

  For those that were more neutral, the principal reason was that:
    - Energy-efficient design already is standard practice in the municipality without being a code requirement. Complying with a codified design guideline for energy efficiency will result in additional code enforcement and compliance reporting costs.
- Primary barriers to implementing energy efficiency projects are additional capital cost of energy-efficient solutions given limited municipal capital budgets and, a related barrier, the lack of political will to support solutions that are more expensive. Suggested pathways to address the first barrier include the following:
    - Enabling the SRF, the primary funding source for wastewater system projects, to reduce the capital cost of energy efficiency projects with incentives such as low-interest loans or principal forgiveness. The resulting improved project life-cycle cost is more palatable politically to the municipality.
    - Changing low-bid procurement rules to recognize more expensive energy efficiency projects.

4.1.2 State and Utility WWTF Energy Efficiency Programs

Eight exemplary programs showed substantial commonality in best practices. Key features of these programs include the following:

- Implementing energy efficiency measures using a performance-based contract with the program implementation contractor.
- Maintaining lists of qualified engineering consultants and project implementation contractors.
- Supporting a variety of marketing and outreach activities (e.g., case studies, conferences, and workshops) to spread the word about the program and to sell the concept and benefits of energy-efficient design.
- Providing project screening services that include site visits; leveraging project screening services provided by the implementation contractor with self-evaluation tools distributed to participants; and promoting incrementally scalable system designs.
- Providing personalized project facilitation services throughout the project to ensure that it moves to completion and considering providing more program assistance to smaller wastewater systems, which have fewer financial and staff resources than larger wastewater systems have to commit to projects.
- Providing access to independent funding sources.
- Providing financial incentives such as cost-share grants to develop energy efficiency projects.
- Presenting nonenergy benefits of energy efficiency projects (e.g., reduced green house gas emissions and improved sustainability) to municipal decision makers.
- Streamlining the funding process so that it does not place a burden on applicants.
- Developing strategies to ensure that investments in energy-efficient equipment and operations are not reversed at a future date.
- Continuing to respond to participants' needs as the program grows; refining program services and providing new energy-efficient technology alternatives as facility operator requirements and the energy-efficient technology market evolve; and learning from experience, which has shown that program participants sell the program to others and continue to use the program if projects address specific site needs with relevant services and solutions that improve their operations cost-effectively.

Table 3-4 presents a framework that incorporates these features.

4.2 Recommendations

- Circulate draft energy efficiency and energy management language to a wider audience and refine it.
- Promote acceptance of energy efficiency and energy management language through involvement of standards organizations and professional societies, such as the Water Environment Federation (WEF), the National Association of Regulatory Utility Commissioners, the Consortium for Energy Efficiency, the ACEEE, the Alliance to Save Energy, and Society of American Value Engineers (SAVE) International.

- Engage state organizations in developing statewide WWTF energy reduction programs through channels such as the National Association of State Energy Officials.
- Provide individuals with experience in the exemplary programs to advise other states and agencies on how to implement statewide energy reduction programs for the wastewater treatment sector.
- Visioning and Planning Activities: Stakeholders can pursue the following recommended activities to advance energy efficiency in wastewater treatment systems. These activities continue in directions suggested by the analysis of state energy reduction program best practices and the analysis that resulted in draft design guideline and standard wording:
  - Engage governing bodies to acknowledge the value available to them from energy-efficient operation of their utilities.
  - Continue to develop case studies that provide insight into the value of energy-efficient exemplary projects. In such case studies and in other energy efficiency publications, show through sensitivity analyses how energy price increases result in more cost-effective energy efficiency projects.
  - Develop an outreach program to train facility designers to create designs that are energy efficient throughout a facility's operating life.
  - Develop a financial incentive program to jump-start energy efficiency projects.
  - Expand market transformation efforts to all parties involved with the wastewater industry.
  - Develop procurement resources that communities can include in their procurement procedures to ensure that wastewater treatment system projects consider energy-efficient design alternatives. For example, the Consortium for Energy Efficiency has developed a memorandum, "Draft RFQ/RFP Guidance for Water/Wastewater Projects", that offers municipal procurement officials sample energy efficiency language they can incorporate into project request documents.
  - Develop and provide energy consumption standards (e.g., kilowatt- or cubic feet-per minute for an energy-efficient aeration blower) that communities can incorporate into their project design guidelines.

REFERENCES

Bidwell, John (QuEST), Jennifer Fagan (Itron), Todd Amundson (BPA), "Water-Wastewater Program Best Practices: Lessons from the Field", presented at ACEEE 2008 Summer Study on Energy Efficiency in Buildings, Pacific Grove, CA, August 2008.

Carns, Keith. "Bringing Energy Efficiency to the Water and Wastewater Industry: How Do We Get There?" *Proceedings of the Water Environment Federation*. WEFTEC **2005**: Sessions 91-100.

Commonwealth of Pennsylvania. *Domestic Wastewater Facilities Manual*. Document ID 362-0300-001, October 1, 1997.

Commonwealth of Virginia, *Sewage Collection and Treatment Regulations* (9VAC25-790), authorized under the Code of Virginia 1950, as amended, Title 62.1, Section 62.1-44.19, http://www.deq.state.va.us/wastewater/regulations.html.

Kezar, Kristi (SAIC), Joseph Cantwell (SAIC), "Lessons in Designing and Delivering Energy Efficiency Programs for the Water and Wastewater Industry", white paper.

Massachusetts Department of Environmental Protection, Massachusetts Energy Management Pilot, http://www.mass.gov/dep/water/wastewater/empilot.htm, September 2009.

Michael DiBara, Massachusetts Energy Management Pilot for Water & Wastewater Plants, EPA New England National Sustainable Infrastructure Meeting, November 6, 2008, MA EM Pilot EPA Sustainability_11-06-08_2 (shrunk).ppt.

New England Interstate Water Pollution Control Commission, *2007 Annual Report,* http://www.neiwpcc.org/neiwpcc_docs/AR2007.pdf.

New England Interstate Water Pollution Control Commission, *TR 16 Guides for the Design of Wastewater Treatment Works,* http://www.neiwpcc.org/tr16guides.asp, 1998.

New York Energy $mart^{sm} Program, Quarterly Evaluation and Status Report, Quarter Ending September 30, 2008, Final Report, http://www.nyserda.org/publications/3rd%20Quarter%202008%20Report.pdf, November 2008.

State of Washington, *Criteria for Sewage Works Design* (Orange Book), http://www.ecy.wa.gov/biblio/9837.html, January 1978 (revised August 2008).

Ten States – Wastewater Committee of the Great Lakes – Upper Mississippi River Board of State and Provincial Public Health and Environmental Managers, *Recommended Standards for Wastewater Facilities,* http://10statesstandards.com/wastewaterstandards.html, 2004.

APPENDIX A

DATABASE OF DESIGN GUIDELINES FOR THE 50 STATES AND FEDERAL GOVERNMENT

A.1 Design Guideline References

State	Agency	Regulatory Role	Design Guideline Citation	Energy Management Requirement Content in Current Design Guideline	Exemplary or Best-Practice Energy Management Content in Current Design Guideline (or in associated State Energy Program [e.g., NY or WI], but not yet reflected in WWTF design guideline issued by permitting agency)
Alabama	Alabama Department of Environmental Management, Water Division, Municipal Branch		None. Alabama does not have a design guideline document.	Per telephone conversations with Ms. Dean and Mr. Daley (7/11/08), Alabama has no guidelines for WWTF design. The staff reviews designs. CWSRF program requires applicants to submit four design alternatives, including a "no-action" alternative. The no-action alternative reflects negative environmental impact and loss of chance to save more water if no change is implemented. The WWTF submits an Environmental Information Document (not	

State	Agency	Regulatory Role	Design Guideline Citation	Energy Management Requirement Content in Current Design Guideline	Exemplary or Best-Practice Energy Management Content in Current Design Guideline (or in associated State Energy Program [e.g., NY or WI], but not yet reflected in WWTF design guideline issued by permitting agency)
				available on the web, and apparently, the format is left to the applicant). The CWSRF program does have an application form available on the web (see link above). Alabama does follow guidance of the Ten States Standards.	
Alabama					
Alaska	Alaska Department of Environmental Conservation, Division of Water, Wastewater Discharge Authorization Programs		18 AAC 72 http://www.dec.state.ak.us/regulations/pdfs/18%20AAC%2072.pdf In particular, 18 ACC 72.245 (Treatment works) 18 ACC 72.220 (Plan review)	No. However, given cold climate and related environmental considerations (e.g., tundra and permafrost), design of wastewater disposal systems includes conduit insulation and cold weather engineering practices. Under the state's Municipal Grants and Loans Program, a WWTF funding source, it is also common	No. However, the Alaska Energy Authority has been studying energy efficiency and alternative energy source options. Additionally, the governor is addressing energy efficiency through two subcabinet activities on climate change and rural issues.

State	Agency	Regulatory Role	Design Guideline Citation	Energy Management Requirement Content in Current Design Guideline	Exemplary or Best-Practice Energy Management Content in Current Design Guideline (or in associated State Energy Program [e.g., NY or WI], but not yet reflected in WWTF design guideline issued by permitting agency)
				practice to consider energy efficiency, especially as a way to offset high-energy costs in rural areas. Another energy efficiency driver in Alaska is the difficulty in getting diesel fuel to rural areas during the long winter season. Shipping is restricted to ice-free conditions; air-freighted fuel is an expensive alternative.	
Arizona	Arizona Department of Environmental Quality, Water Quality Division (WQD)				
Arkansas	State of Arkansas Department of Environmental Quality, Water Division, Permits Branch	Issues permits for WWTF construction	Ten States Standards and Manual of Practices (MOP)	No.	
California	State Water Resources Control Board		No design guideline for WWTF. California's Water Code prohibits the State Water Board from specifying the manner of compliance with discharge permits.	No. California's CWSRF program does accept energy efficiency projects, either as the sole project objective or as part of a project with broader scope. CWSRF does	

State	Agency	Regulatory Role	Design Guideline Citation	Energy Management Requirement Content in Current Design Guideline	Exemplary or Best-Practice Energy Management Content in Current Design Guideline (or in associated State Energy Program [e.g., NY or WI], but not yet reflected in WWTF design guideline issued by permitting agency)
				require life-cycle cost analysis, and energy costs are embedded in such analyses.	
Colorado	Water Quality Control Division (WQCD)		Referenced website contains policies and design criteria for the siting and construction of wastewater facilities in Colorado.		
Colorado	WQCD				
Connecticut	Connecticut Department of Environmental Protection		TR-16, 1998 version by NEIWPCC	No.	Northeast Utilities established a group to provide grants. Contact: James Motta, Supervisor in Conservation and Load Management Program, mottaj@nu.com
Delaware	Department of Natural Resources and Environmental Control, Division of Water Resources and Surface Water Discharges Section	Issues permits for WWTF construction			
District of Columbia	District of Columbia Water and Sewer Authority, Capital Improvement Program				

State	Agency	Regulatory Role	Design Guideline Citation	Energy Management Requirement Content in Current Design Guideline	Exemplary or Best-Practice Energy Management Content in Current Design Guideline (or in associated State Energy Program [e.g., NY or WI], but not yet reflected in WWTF design guideline issued by permitting agency)
Florida	Florida Department of Environmental Protection, Office of Wastewater Management, Domestic Wastewater Section		Rule 62-600 Florida Administrative Code/Domestic Wastewater Facilities– 62-600.400 (Design Requirements)(http://www.dep.state.fl.us/legal/Rules/wastewater/62-600.pdf); Guide to Permitting Wastewater Facilities or Activities under Chapter 62-620, F.A.C. (http://www.dep.state.fl.us/water/wastewater/dom/docs/wwguide.pdf)	No.	No.
Georgia	Georgia Department of Natural Resources, Environmental Protection Division, Watershed Protection Branch		Per 7/21/08 e-mail, Bob Scott, Program Manager, Engineering and Technical Support Program, indicated the department does not have guidance on energy efficiency or management best practices. Focus is on the technical side. Suggested Frances Carpenter as contact person in this program.		
Hawaii	Hawaii Department of Land and Natural Resources, Engineering Division				

State	Agency	Regulatory Role	Design Guideline Citation	Energy Management Requirement Content in Current Design Guideline	Exemplary or Best-Practice Energy Management Content in Current Design Guideline (or in associated State Energy Program [e.g., NY or WI], but not yet reflected in WWTF design guideline issued by permitting agency)
Idaho	Department of Environmental Quality, Water Quality Division, Wastewater Program		http://adm.idaho.gov/adminrules/rules/idapa58/0116.pdf	No.	No.
Illinois	Illinois EPA		http://www.epa.state.il.us/water/rules-regulation.html	No.	Electric utilities, such as Exelon and Ameren, fund energy efficiency projects for WWTFs.
Indiana	Indiana Department of Environmental Management		Ten States Standards and MOPs	No.	
Iowa	Iowa Department of Natural Resources (DNR), Water Quality Bureau, Wastewater Construction Section	Wastewater Construction Section issues construction permits for municipal and industrial facilities; administers Wastewater State Revolving Fund Loan Program.	Iowa Wastewater Design Standards http://www.iowadnr.com/water/wastewater/index.html, Ten States Standards http://10statesstandards.com/, Iowa Wastewater Facilities Design Standards http://www.iowadnr.com/water/wastewater/downloads.html	No.	No.
Iowa	Iowa DNR, Water Quality Bureau, Wastewater Construction Section				
Kansas	Bureau of Water Division of Environment, Kansas Department of Health and Environment (KDHE)		Minimum Standards of Design	Minimum Standards of Design last updated in 1978; many parts are out of date. The document does not address energy	Continue to encourage the use of lagoon technology for treatment, as no "man-made" energy input is required; also encourage all "mechanical" treatment processes to conduct energy efficiency reviews, but no written guidance, yet.

State	Agency	Regulatory Role	Design Guideline Citation	Energy Management Requirement Content in Current Design Guideline	Exemplary or Best-Practice Energy Management Content in Current Design Guideline (or in associated State Energy Program [e.g., NY or WI], but not yet reflected in WWTF design guideline issued by permitting agency)
				efficiency of designs or benchmarks of energy-efficient operation, but is directed toward quality of construction and process efficiency to remove pollutants.	
Kentucky	Kentucky Department of Environmental Protection, Division of Water, Facilities Construction Branch, Municipal Planning Section	Planning of municipal wastewater collection and treatment systems			
Louisiana	Louisiana Department of Environmental Quality (DEQ), Office of Environmental Services, Water Permits Division	Municipal and general water permits			
Louisiana	Louisiana DEQ, Office of Environmental Services, Waste Permits Division, Clean Water State Revolving Fund	SRF program for construction of wastewater treatment systems			No.
Louisiana	Louisiana Department of Health and Hospitals (DHH), Office of Public Health (OPH), Center	Wastewater treatment facility review			

State	Agency	Regulatory Role	Design Guideline Citation	Energy Management Requirement Content in Current Design Guideline	Exemplary or Best-Practice Energy Management Content in Current Design Guideline (or in associated State Energy Program [e.g., NY or WI], but not yet reflected in WWTF design guideline issued by permitting agency)
	for Environmental Health, Engineering Services, Safe Drinking Water Program				
Maine	Maine Department of Environmental Protection		TR-16 and statewide building code follows American Society of Heating, Refrigerating and Air Conditioning Engineers (ASHRAE) guidelines.	No, not WWTF-specific. Energy management is addressed in value engineering studies.	Efficiency Maine, www.efficiencymaine.com, provides funding
Maryland	Maryland Department of Environment, Water Management Administration				
Massachusetts	Massachusetts Department of Environmental Protection (MassDEP)		TR 16 http://www.neiwpcc.org/tr16guides.asp	No. However, TR 16 update is in process. In September 2008, the Executive Committee of NEIWPCC voted to update the TR-16 Wastewater Design manual to incorporate energy efficiency.	Yes. Two targeted efforts are designed to improve energy efficiencies at wastewater treatment facilities: 1) MA Energy Management Pilot for Water and Wastewater: Launched in December 2007, MassDEP has collaborated with every major electric and gas utility provider in the state to assess, advance, and implement energy saving improvements at these pilot plants. http://www.mass.gov/dep/public/publications/0108ener.htm. Contact: Michael DiBara, Project Manager, michael.dibara@state.ma.us 2) MA SRF: In 2007, MassDEP expanded its eligibility for

State	Agency	Regulatory Role	Design Guideline Citation	Energy Management Requirement Content in Current Design Guideline	Exemplary or Best-Practice Energy Management Content in Current Design Guideline (or in associated State Energy Program [e.g., NY or WI], but not yet reflected in WWTF design guideline issued by permitting agency)
					energy-saving projects in facility upgrades at wastewater and drinking water plants. Contact: Steven McCurdy, http://www.mass.gov/dep/energy.htm.
Michigan	Michigan DEQ		Ten States Standards and MOPs	No.	No.
Minnesota	Minnesota Pollution Control Agency		Ten States Standards and MOPs. Additionally, The State of Minnesota Sustainable Building Guidelines version 2.0 <http://www.msbg.umn.edu/> 2008 Minnesota Statutes 16B.325 (Sustainable Building Guidelines) website https://www.revisor.leg.state.mn.us/statutes/?id=16B.325 address building energy efficiency, but not WWTFs.	No. Sustainable Building Guidelines <http://www.msbg.umn.edu/energy.html> Energy & Atmosphere Section – reduce energy use by 30%, provide building performance data for benchmarking activities; reduce plug loads and process energy through energy-smart purchasing practices; encourage the consideration of power usage from renewable energy and cleaner generation systems, whether generated on site or	1) Minnesota Sustainable Building Guideline; contact name: Richard Strong; stron081@umn.edu; http://www.msbg.umn.edu/ 2) Minnesota Conservation Improvement Program; Bruce Nelson; bruce.nelson@state.mn.us; http://www.commerce.state.mn.us/http://www.state.mn.us/portal/mn/jsp/content.do?subchannel=-536881736&id=-536881351&agency=Commerce

State	Agency	Regulatory Role	Design Guideline Citation	Energy Management Requirement Content in Current Design Guideline	Exemplary or Best-Practice Energy Management Content in Current Design Guideline (or in associated State Energy Program [e.g., NY or WI], but not yet reflected in WWTF design guideline issued by permitting agency)
				purchased from off-site "green power" generated in Minnesota; encourage the balanced consideration of global warming potential, ozone depletion, and atmospheric lifetime in selecting refrigerants; help ensure that long-term operations meet or exceed original design operating parameters. Sustainable Buildings Guidelines do not address WWTFs.	
Mississippi	Mississippi Department of Environmental Quality		Not referenced in e-mail.		
Missouri	Missouri Department of Natural Resources, Water Protection Program				
Montana	Montana DEQ, Permitting and Compliance Division, Water Protection Bureau, Water Quality Discharge Permits Section		DEQ-4 and DEQ-2: http://www.deq.mt.gov/wqinfo/Circulars.asp	No.	No.

State	Agency	Regulatory Role	Design Guideline Citation	Energy Management Requirement Content in Current Design Guideline	Exemplary or Best-Practice Energy Management Content in Current Design Guideline (or in associated State Energy Program [e.g., NY or WI], but not yet reflected in WWTF design guideline issued by permitting agency)
Nebraska	Nebraska DEQ, Water Quality Division, Wastewater Section, Wastewater Engineering Section		Title 123 (Nebraska Department of Environmental Quality); Chapter 5 (Design Standards and Specification)	No.	
Nevada	Nevada Division of Environmental Protection, Bureau of Water Pollution Control, Technical Services Branch		Various design guidance documents (e.g., WTS-5 [design of wastewater treatment ponds], WTS-14 [pumping station design]) (http://www.ndep.nv.gov/bwpc/fact01.htm); also use Ten States Standards and WEF Design Manuals	No. However, while not a requirement, the staff evaluates energy management as part of the SRF facility planning process.	No. Also, no programs in either the Nevada State Office of Energy or Nevada Division of Environmental Protection.
New Hampshire	New Hampshire (NH) Department of Environmental Services, Wastewater Engineering Bureau, Design Review Section		NH Code of Administrative Rules, Chapter Env-Wq 700 – Standards of Design and Construction for Sewage and Wastewater Treatment Facilities. http://des.nh.gov/organization/commissioner/legal/rules/documents/env-wq700.pdf	No.	Amy Ignatius, Director, NH Office of Energy and Planning, OEPinfo@nh.gov, www.nh.gov.oep
New Jersey	New Jersey Department of Environmental Protection		www.nj.gov/dep/dwq/714a.htm	No.	1) Clean Water State Revolving Funds, www.nj.gov/dep/dwq/mface.htm; 2) board of public utilities.
New Mexico	New Mexico Environment Department, Water and Wastewater Infrastructure Development Division,		http://www.nmenv.state.nm.us/cpb/Wastewater%20Guidelines%20%20Rev.%202003.pdf	No.	No.

State	Agency	Regulatory Role	Design Guideline Citation	Energy Management Requirement Content in Current Design Guideline	Exemplary or Best-Practice Energy Management Content in Current Design Guideline (or in associated State Energy Program [e.g., NY or WI], but not yet reflected in WWTF design guideline issued by permitting agency)
	Construction Programs Bureau				
New York	New York Department of Environmental Conservation		Ten States Standards	No.	1) NYSERDA provides funding for energy efficiency upgrades. http://www.nyserda.org/ 2) Environmental Facilities Corporation is another grant and funding agency. http://www.nysefc.org/home/index.asp
North Carolina	North Carolina Department of Environmental and Natural Resources, Division of Water Quality, Construction Grants and Loans Section		WWTF Authorization to Construct Guidance, http://www.nccgl.net/Engineering/ATCguidance.pdf		
North Dakota	North Dakota Department of Health, Environmental Health Section, Division of Municipal Facilities, Clean Water State Revolving Fund Program				
Ohio	Ohio EPA		Ten States Standards	No.	No.
Oklahoma	State of Oklahoma DEQ, Water Quality Division		Title 252 (DEQ) Chapter 656 (Water Pollution Control Facility Construction Standards) (http://www.deq.state.ok.us/rules/656.pdf)		

State	Agency	Regulatory Role	Design Guideline Citation	Energy Management Requirement Content in Current Design Guideline	Exemplary or Best-Practice Energy Management Content in Current Design Guideline (or in associated State Energy Program [e.g., NY or WI], but not yet reflected in WWTF design guideline issued by permitting agency)
Oregon	Oregon DEQ, Water Quality Division, Standards and Assessments		None.	No.	No.
Pennsylvania	Commonwealth of Pennsylvania Department of Environmental Protection, Bureau of Water Standards and Facility Regulation	Issues permits for WWTF construction; monitors compliance	Domestic Wastewater Facilities Manual, Document ID 362-0300-001, October 1, 1997.	Section 43 (Design): 43.1 (Type of Treatment) – Cites "energy requirements" among "other considerations" for design. Section 82 (Activated Sludge):82.113 (Energy Requirements)– recognizes aeration as a major energy demand; requires energy costs to be carefully evaluated; requires activated sludge process design to include "Capability of energy usage phase-down, while still maintaining process viability, both under normal and emergency energy availability conditions". Section 82.332.e	

State	Agency	Regulatory Role	Design Guideline Citation	Energy Management Requirement Content in Current Design Guideline	Exemplary or Best-Practice Energy Management Content in Current Design Guideline (or in associated State Energy Program [e.g., NY or WI], but not yet reflected in WWTF design guideline issued by permitting agency)
				(Diffused Air Systems) – requires blowers to be in multiple units so that maximum air demand can be met with the largest unit out of service; blowers must be able to vary volume of air in response to plant's load demand; "use of timers to conserve energy is acceptable, provided necessary minimum dissolved oxygen level is maintained in the aeration tank". Section 83 (RBCs): 83.3 (Design Considerations): 83.31 (Process Loading Reliability): 83.31.n – indicates "use of high-efficiency motors for mechanically driven RBC units to lower the energy consumption should be considered".	

State	Agency	Regulatory Role	Design Guideline Citation	Energy Management Requirement Content in Current Design Guideline	Exemplary or Best-Practice Energy Management Content in Current Design Guideline (or in associated State Energy Program [e.g., NY or WI], but not yet reflected in WWTF design guideline issued by permitting agency)
Rhode Island	Rhode Island (RI) Department of Environmental Management, Division of Water Resources, Planning and Design Section		TR-16	No.	RI office of Energy Resources - http://www.energy.ri.gov/programs/renewable.php
South Carolina	South Carolina Department of Health and Environmental Control	WWTF design review	SCDHEC Standards for Wastewater Facility Construction (R.61-67) (http://www.scdhec.gov/environment/water/regs/r61-67.doc); Guidance (http://www.scdhec.gov/environment/water/docs/wwtp_new.pdf)	No.	
South Dakota	South Dakota Department of Environment and Natural Resources				
Tennessee	Tennessee Department of Environment and Conservation				
Texas	Texas Commission on Environmental Quality (CEQ)		http://www.tceq.state.tx.us/permitting/water_quality/wastewater/plans/WW_design_requirements.html		
Texas	Texas CEQ		Documents sent include Chapter 317 and draft Chapter 217 design criteria for wastewater treatment system.	No.	

State	Agency	Regulatory Role	Design Guideline Citation	Energy Management Requirement Content in Current Design Guideline	Exemplary or Best-Practice Energy Management Content in Current Design Guideline (or in associated State Energy Program [e.g., NY or WI], but not yet reflected in WWTF design guideline issued by permitting agency)
Utah	Utah DEQ, Division of Water Quality, Engineering Section		http://www.rules.utah.gov/publicat/code/r317/r317-003.htm	No.	No, although the department informs the public regarding the energy efficiency of treatment alternatives when spending SRF funds on projects.
Vermont	Vermont Department of Natural Resources		Ten States Standards and TR-16	No.	The state building code has energy efficiency standards that apply to wastewater facilities.
Virginia	Virginia DEQ, Water Quality Division, Office of Wastewater Engineering	Design approval; issues permits for WWTF construction; provides technical assistance	Sewage Collection and Treatment Regulations (9 VAC 25-790), authorized under the Code of Virginia 1950, as amended, Title 62.1, Section 62.1-44.19 (http://www.deq.state.va.us/wastewater/regulations.html)	9VAC25-790-550 (Anaerobic digestion) F. Energy Control "If practical, digesters should be constructed above the water table and should be suitably insulated to minimize heat loss. The use of digester gas as a heating fuel source is encouraged". 9VAC25-790-560 (Aerobic sludge digestion) C. Loadings "2. Energy input requirements for mixing should be in the range of 0.5 to 1.5 horsepower per 1,000 cubic feet, where mechanical aerators are utilized, and 20 to 30 standard cubic feet per minute per 1,000 cubic feet of aeration tank,	

State	Agency	Regulatory Role	Design Guideline Citation	Energy Management Requirement Content in Current Design Guideline	Exemplary or Best-Practice Energy Management Content in Current Design Guideline (or in associated State Energy Program [e.g., NY or WI], but not yet reflected in WWTF design guideline issued by permitting agency)
				where air mixing is utilized". 9VAC25-790-770 (UV) "E. Dose control. For treatment works with a design average daily flow of one mgd or more, UV system design should include a control system to turn appropriate lamps on or off in order to conserve energy".	
Washington	Washington Department of Ecology		State of Washington, Department of Ecology, Criteria for Sewage Works Design (Orange Book), http://www.ecy.wa.gov/biblio/9837.html, January 1978 (revised August 2008).	Requirements for Engineering Report and Facility Plan (see Design Guideline Citation, Table G1-1, p. G1-18) states that the facility plan alternatives analysis is subject to the same requirements as the engineering report "plus description of innovative and alternative technologies (that is, those saving energy and nonconventional treatment [land application, etc.])".	

State	Agency	Regulatory Role	Design Guideline Citation	Energy Management Requirement Content in Current Design Guideline	Exemplary or Best-Practice Energy Management Content in Current Design Guideline (or in associated State Energy Program [e.g., NY or WI], but not yet reflected in WWTF design guideline issued by permitting agency)
West Virginia	West Virginia Water Development Authority	Provides financing for design, construction, improvement, and acquisition of wastewater and water facilities to local government agencies			
Wisconsin	Wisconsin DNR	Design approval	Chapter NR 110 Wisconsin Administrative Code (http://dnr.wi.gov/org/water/wm/glwsp/facilities/coderef.htm, and select the NR 110 regulation from the list of regulations. NR 110 addresses design standards for municipal wastewater facilities.)	No requirements that specifically address any aspect of energy efficiency. Wisconsin does have a requirement that proposed wastewater facilities must be demonstrated to be cost-effective; cost analyses must be conducted during a 20-year planning period and account for electrical power costs.	

State	Agency	Regulatory Role	Design Guideline Citation	Energy Management Requirement Content in Current Design Guideline	Exemplary or Best-Practice Energy Management Content in Current Design Guideline (or in associated State Energy Program [e.g., NY or WI], but not yet reflected in WWTF design guideline issued by permitting agency)
Wyoming	Wyoming Department of Environmental Quality, Water Quality Division	Design approval; issues permits for WWTF construction.	Water Quality Rules and Regulations, Chapter IX (Design and Construction Standards for Sewerage Systems, Treatment Works, Disposal Systems or Other Facilities Capable of Causing or Contributing to Pollution and Mobile Home Park and Campground Sewerage and Public Water Supply Distribution Systems), Parts A and B, http://deq.state.wy.us/wqd/WQDrules/Chapter_11.pdf	No.	
Federal	U.S. EPA				

A.2 Design Guideline Contact Information

State	Agency	Contact Information		Mailing Address	Web Address
		Name	E-mail		
Alabama	Alabama Department of Environmental Management (ADEM), Water Division, Municipal Branch	Glenda Dean, Branch Chief, Permitting, Municipal Branch	gld@adem.state.al.us		http://www.adem.state.al.us/WaterDivision/Municipal/MunicipalWater.htm
Alabama		James Daley, Manager, ADEM CWSRF Loan Program	jwd@adem.state.al.us		
Alaska	Alaska Department of Environmental Conservation, Division of Water, Wastewater Discharge Authorization Programs	Sharmon M. Stambaugh, Program Manager	sharmon.stambaugh@alaska.gov	Wastewater Discharge Authorization Programs, Alaska Department of Environmental Conservation Division of Water 555 Cordova Street Anchorage AK 99501	http://www.dec.state.ak.us/water/wwdp/index.htm
Arizona	Arizona Department of Environmental Quality, Water Quality Division	Kwame Agyare, PE, Engineering Review Section Manager	ka2@azdeq.gov		http://www.azsos.gov/public_services/Title_18/18-09.htm
Arkansas	State of Arkansas Department of Environmental Quality, Water Division, Permits Branch	John Bailey, Permits Branch Manager	bailey@adeq.state.ar.us		http://www.adeq.state.ar.us/water/branch_permits/individual_permits/wastewater/default.htm
California	State Water Resources Control Board	Todd Thompson, PE, On-Site Treatment Systems Specialist	tthompson@waterboards.ca.gov	Division of Water Quality	http://www.waterboards.ca.gov/water_issues/programs/#wastewater
California	State Water Resources Control Board	Christopher Stevens, Chief, CWSRF and Special Programs Section	cstevens@waterboards.ca.gov	CWSRF Program	http://www.waterboards.ca.gov/water_issues/programs/grants_loans/srf/index.shtml#contact

State	Agency	Contact Information		Mailing Address	Web Address
		Name	E-mail		
Colorado	Water Quality Control Division (WQCD)	Jennifer Miller, Manager, Denver Field Office, WQCD Engineering Section	jennifer.miller@state.co.us		www.cdphe.state.co.us/wq/engineering/techhom.html
Colorado	WQCD	Betsy Beaver, Facility-Operator Program, WQCD	betsy.beaver@state.co.us		
Connecticut	Connecticut Department of Environmental Protection	William Hogan, Engineer of Water Pollution Control	william.hogan@po.state.ct.us		http://www.ct.gov/dep/site/default.asp
Delaware	Department of Natural Resources and Environmental Control (DNREC), Division of Water Resources, Surface Water Discharges Section	Section Manager: R. Peder Hansen, PE	Peder.Hansen@state.de.us	DNREC, Division of Water Resources Surface Water Discharges Section 89 Kings Highway Dover, DE 19901	http://www.wr.dnrec.delaware.gov/Services/Pages/SurfaceWaterDischarges.aspx
District of Columbia	District of Columbia Water and Sewer Authority, Capital Improvement Program	Steven Hamburger, Manager of Specifications and Estimating	steven_hamburger@dcwasa.com		http://www.dcwasa.com/about/facilities.cfm
Florida	Florida Department of Environmental Protection, Office of Wastewater Management, Domestic Wastewater Section	Richard L. Addison, PE, Domestic Wastewater Section	richard.addison@dep.state.fl.us	2600 Blair Stone Road MS 3540 Tallahassee, FL 32399-2400	http://www.dep.state.fl.us/mainpage/programs/wastewater.htm

State	Agency	Contact Information		Mailing Address	Web Address
		Name	E-mail		
Georgia	Georgia Department of Natural Resources, Environmental Protection Division, Watershed Protection Branch	Frances Carpenter, Watershed Protection Branch, Engineering and Technical Support Program, Engineering Unit 2	frances.carpenter@dnr.state.ga.us		http://www.georgiaepd.org/Documents/index_water.html
Hawaii	Hawaii Department of Land and Natural Resources, Engineering Division	Eric T. Hirano, PE, Chief Engineer	eric.t.hirano@hawaii.gov		http://hawaii.gov/dlnr/
Idaho	Department of Environmental Quality, Water Quality Division, Wastewater Program	Richard Huddleston, PE, Wastewater Manager	richard.huddleston@deq.idaho.gov		http://www.deq.idaho.gov/water/prog_issues/waste_water/overview.cfm
Illinois	Illinois EPA	Dennis McMurray	dennis.mcmurray@illinois.gov		http://www.epa.state.il.us/water/
Indiana	Indiana Department of Environmental Management	Don Worley, Senior Environmental Manager, Facility Construction and Engineering Support Section	dworley@idem.in.gov		http://www.in.gov/idem/
Iowa	Iowa Department of Natural Resources, Water Quality Bureau, Wastewater Construction Section	Wayne Farrand, PE, Section Supervisor, Wastewater Construction	wayne.farrand@dnr.iowa.gov	502 East 9th Street Des Moines, IA 50319-0034	www.iowadnr.gov/water/index.html
Iowa	Iowa Department of Natural Resources, Water Quality Bureau, Wastewater	Satya Chennupati, Environmental Engineer Senior, Wastewater Construction	satya.chennupati@dnr.iowa.gov	502 E. 9th Street Des Moines, IA 50319-0034	www.iowadnr.gov/water/index.html

State	Agency	Contact Information		Mailing Address	Web Address
		Name	E-mail		
	Construction Section				
Kansas	Bureau of Water Division of Environment, KDHE	Rodney R. Geisler, PE, Chief, Municipal Programs Section	RGeisler@kdhe.state.ks.us	Municipal Programs Section, Bureau of Water, Division of Environment, KDHE 1000 SW Jackson, Suite 420 Topeka, KS 66612-1367	http://www.kdheks.gov/muni/index.htm
Kentucky	Kentucky Department of Environmental Protection, Division of Water, Facilities Construction Branch, Municipal Planning Section	Hamid Beykezadeh	hamid.beykezadeh@ky.gov not a good e-mail	Division of Water 200 Fair Oaks Lane, Fourth Floor Frankfort, KY 40601	http://water.ky.gov/wastewater/
Louisiana	Louisiana DEQ, Office of Environmental Services, Water Permits Division	Melvin "Mitch" Mitchell, Division Administrator	mitch.mitchell@la.gov	Office of Environmental Services P.O. Box 4313 Baton Rouge, LA 70821-4313	http://www.deq.louisiana.gov/portal/tabid/2229/Default.aspx
Louisiana	Louisiana DEQ, Office of Environmental Services, Waste Permits Division, Clean Water State Revolving Fund	Tom Griggs, Engineer Manager	Tom.Griggs@la.gov	P.O. Box 4301 Baton Rouge, LA 70821-4301	http://www.deq.louisiana.gov/portal/tabid/2229/Default.aspx
Louisiana	Louisiana DHH , OPH, Center for Environmental Health, Engineering Services, Safe Drinking Water Program	Karen Irion, PE, Chief Engineer	Kirion@dhh.la.gov	DHH-OPH-Center for Environmental Health P.O. Box 4489 Baton Rouge, LA 70821-4489	http://www.dhh.louisiana.gov/offices/?ID=204
Maine	Maine Department of Environmental Protection	Dick Darling, Municipal Pollution Prevention, Senior	dick.darling@main.gov	Bureau of Land and Water Quality	www.maine.gov/dep

State	Agency	Contact Information		Mailing Address	Web Address
		Name	E-mail		
		Environmental Engineer			
Maryland	Maryland Department of Environment, Water Management Administration	Dr. Ta-Shon Yu, Water Quality Infrastructure Program	tyu@mde.state.md.us	Maryland Department of the Environment 1800 Washington Boulevard Baltimore, MD 21230 Attention: Permit Coordinator, Water Quality Infrastructure Program, Water Management Administration	http://www.mde.state.md.us/Permits/WaterManagementPermits/index.asp
Massachusetts	Department of Environmental Protection	Steven McCurdy, Director, Division of Municipal Services	steven.mccurdy@state.ma.us		http://www.mass.gov/dep
Michigan	Michigan DEQ	Dave Porter	porterd2@michigan.gov		http://www.michigan.gov/deq/
Minnesota	Minnesota Pollution Control Agency	Dave Sahli	david.sahli@state.mn.us		www.pca.state.mn.us
Mississippi	Mississippi Department of Environmental Quality	Jeff Pittman	jeff_pittman@deq.state.ms.us	Water Pollution Control (Clean Water) Revolving Loan Fund (WPCRLF) Program	http://www.deq.state.ms.us/MDEQ.nsf/page/SRF_Water_PC_RLP?OpenDocument
Missouri	Missouri Department of Natural Resources, Water Protection Program	Refaat Mefrakis, Section Chief	refaat.mefrakis@dnr.mo.gov		http://www.dnr.mo.gov/
Montana	Montana DEQ, Permitting and Compliance Division, Water Protection Bureau, Water Quality Discharge Permits Section	John McDunn, Environmental Engineer	jmcdunn@mt.gov		http://www.deq.state.mt.us/

State	Agency	Contact Information		Mailing Address	Web Address
		Name	E-mail		
Nebraska	Nebraska DEQ, Water Quality Division, Wastewater Section, Wastewater Engineering Section			1200 "N" Street, Suite 400 P.O. Box 98922 Lincoln, Nebraska 68509	http://www.deq.state.ne.us/ (At this site, click Focus on Water, then click Wastewater Construction Permit Program.)
Nevada	Nevada Division of Environmental Protection, Bureau of Water Pollution Control, Technical Services Branch	Joseph L. Maez, PE, engineer in charge of SRF reviews	jmaez@ndep.nv.gov	901 South Stewart Street, Suite 4001 Carson City, Nevada 89701-5249	http://www.ndep.nv.gov/bwpc/index.htm
New Hampshire	New Hampshire Department of Environmental Services, Wastewater Engineering Bureau, Design Review Section	Steve Roberts, Senior Sanitary Engineer	sroberts@des.nh.gov		http://des/nh.gov
New Jersey	New Jersey Department of Environmental Protection	Ted Ovsiew, Staff Engineer	theodore.ovsiew@dep.state.nj.us		http://www.state.nj.us/dep/
New Mexico	New Mexico Environment Department Water and Wastewater Infrastructure Development Division, Construction Programs Bureau	Richard Rose	richard.rose@state.nm.us		http://www.nmenv.state.nm.us/
New York	New York State Department of Environmental Conservation	Cheryle Webber, Section Chief, Wastewater Permits	clwebber@gw.dec.state.ny.us		http://www.dec.ny.gov/
New York	New York State Environmental Services Corporation, Clean Water State Revolving	Tim Burns, Technical Services	burns@nysefc.org		http://www.nysefc.org/

State	Agency	Contact Information		Mailing Address	Web Address
		Name	E-mail		
	Fund				
North Carolina	North Carolina Department of Environmental and Natural Resources, Division of Water Quality, Construction Grants and Loans Section	Daniel M. Blaisdell, PE, Assistant Chief, Engineering Branch, Construction Grants and Loans Section	Daniel.Blaisdell@ncmail.net	NC Division of Water Quality, Construction Grants and Loans Section 1633 Mail Service Center Raleigh, NC 27699-1633 Attention: Daniel M. Blaisdell, PE	http://www.nccgl.net/fap/cwsrf/index.html
North Dakota	North Dakota Department of Health, Environmental Health Section, Division of Municipal Facilities, Clean Water State Revolving Fund Program	Jeff Hauge, PE, Program Manager, Clean Water State Revolving Loan Fund Program	jhauge@nd.gov	North Dakota Department of Health Division of Municipal Facilities 918 East Divide Avenue, 3rd Floor Bismarck, ND 58501-1947	http://www.ndhealth.gov/EHS/
Ohio	Ohio EPA	Daniel Kopec, Environmental Specialist 2	daniel.kopec@epa.state.oh.us		www.epa.state.oh.us/
Oklahoma	State of Oklahoma DEQ, Water Quality Division	Rocky Chen, Construction Permits		Oklahoma DEQ 707 North Robinson, Oklahoma City, OK 73102-6010	http://www.deq.state.ok.us/wqdnew/index.htm
Oregon	Oregon DEQ, Water Quality Division, Standards and Assessments	Jennifer Wigal, Manager	wigal.jennifer@deq.state.or.us		http://www.oregon.gov/DEQ/WQ/
Pennsylvania	Commonwealth of Pennsylvania Department of Environmental Protection, Bureau of Water Standards and Facility Regulation	Dana Aunkst, Director		Bureau of Water Standards and Facility Regulation P.O. Box 8467 Harrisburg, PA 17105-8467	http://www.depweb.state.pa.us/watersupply/cwp/view.asp?a=1450&q=512592

State	Agency	Contact Information		Mailing Address	Web Address
		Name	E-mail		
Rhode Island	Rhode Island Department of Environmental Management, Division of Water Resources, Planning and Design Section	Art Zeman	art.zeman@dem.ri.gov		http://www.state.ri.us/dem/
South Carolina	South Carolina Department of Health and Environmental Control (SCDHEC)	Jeff deBessonet, Director, Water Facilities Permitting Division, SCDHEC	debessjp@dhec.sc.gov	SCDHEC Bureau of Water, 2600 Bull Street, Columbia, SC 29201	http://www.scdhec.gov/environment/water/
South Dakota	South Dakota Department of Environment and Natural Resources	Kelli Buscher, Natural Resources Engineering Director, Surface Water Quality, Pierre	Kelli.Buscher@state.sd.us		http://www.state.sd.us/denr/
Tennessee	Tennessee Department of Environment and Conservation	Emily Kelly, PE,Technical Section Manager, SRF Loan Program	emily.kelly@state.tn.us		http://www.state.tn.us/environment/srf/cwsrf/index.shtml
Texas	Texas CEQ (TCEQ)	L'Oreal Stepney, PE, Division Director	wqap@tceq.state.tx.us	TCEQ, Office of Permitting, Remediation, and Registration Water Quality Division	http://www.tceq.state.tx.us/permitting/water_quality/wastewater/plans/WW_design_requirements.html
Texas	TCEQ	Louis Herrin, PE, contact person for Texas Wastewater Design Criteria	Lherrin@tceq.state.tx.us		www.tceq.state.tx.us
Utah	Utah DEQ, Division of Water Quality, Engineering Section	Ed Macauley, PE, Manager	emacauley@utah.gov		http://www.waterquality.utah.gov/index.htm
Vermont	Vermont Department of Natural Resources	Nopadon Sundarabhaya, Chief Design Section	nope.sundarabhaya@state.vt.us		www.anr.state.vt.us

State	Agency	Contact Information		Mailing Address	Web Address
		Name	E-mail		
Virginia	Virginia DEQ, Water Quality Division, Office of Wastewater Engineering	Marcia Degen, Ph.D., PE,	mjdegen@deq.virginia.gov	DEQ-WCRO 3019 Peters Creek Road Roanoke, VA 24019	http://www.deq.state.va.us/wastewater/
Washington	Washington Department of Ecology	Foroozan Labib	flab461@ecy.wa.gov		http://www.ecy.wa.gov/ecyhome.html
West Virginia	West Virginia Water Development Authority	Chris E. Jarrett, Executive Director	cjarrett@wvwda.org	Water Development Authority 180 Association Drive Charleston, WV 25311-1217	http://www.wvwda.org/
Wisconsin	DNR	Thomas A. Gilbert, PE, Wastewater Facility Planning Coordinator	Tom.Gilbert@dnr.state.wi.us	Wastewater Section, Wisconsin DNR Bureau of Watershed Management P.O. Box 7921 Madison, WI 53707-7921	http://dnr.wi.gov/org/water/wm/glwsp/facilities/munifp.htm
Wyoming	Wyoming Department of Environmental Quality, Water Quality Division	Lou Harmon, Program Manager	lharmo@wyo.gov	Department of Environmental Quality Water Quality Division Water and Wastewater Program 122 West 25th Street Herschler Building, 4th Floor-West Cheyenne, Wyoming 82001	http://deq.state.wy.us/wqd/www/Permitting/Pages/WWS.asp

State	Agency	Contact Information		Mailing Address	Web Address
		Name	E-mail		
Federal	U.S. EPA	Jim Wheeler, U.S. EPA member of the WERF Issue Area Team (IAT) for Optimization of Wastewater and Solids Operations (OWSO) research challenge			Clean Water Revolving Fund history (http://www.epa.gov/owm/cwfinance/privatization.htm)

APPENDIX B

AGENCY MEETING REPORT – DESIGN GUIDELINES ENERGY EFFICIENCY WORDING

Agency Meeting Report – Design Guidelines Energy Efficiency Wording

The following pages report on a discussion among various WWTF organizations on design guidelines energy efficiency wording. Attachment 1 contains written comments on the wording.

Date: July 30, 2009, 4 p.m. EDT–5 p.m. EDT

Attendees:

SAIC – Joe Cantwell (Teleconference Chair and SAIC Program Manager for the WERF energy efficiency tasks), Raffy King (Senior Project Manager), and Bob Lorand (SAIC Deputy Program Manager for WERF energy efficiency tasks)

Green Bay Metropolitan Sewage District (GBMSD) – Dan Busch (Director of Operations)

King County Wastewater Treatment Division (KCWTD, in the Seattle, WA, area) – Roger Browne and Suzanne Schweitzer (Resource Recovery Section Manager)

Metropolitan Water Reclamation District of Greater Chicago (MWRDGC) – Chris Nam, Manickam Annamalai, and Glenn Gottardo

San Antonio Water System (SAWS) – Meg Conner (Director, Treatment) and Dan Titerle

Sheboygan Regional Wastewater Treatment Facility (SRWTF) – Dale Doerr (Superintendent)

Teleconference Discussion Topics:

1. Comments regarding the need for design guideline energy efficiency wording. Would such wording support regional, state, and local initiatives to advance energy efficiency?
2. Comments on design guideline energy efficiency wording.
3. Pathways to incorporating energy efficiency wording into regional, state, and local design guidelines, plus the barriers to incorporating wording and approaches to addressing those barriers.

Draft Design Guideline Energy Efficiency Wording: All projects requiring a facility plan, facility plan amendment, comparison of alternatives, facility modification, or a facility reevaluation for capacity rating shall:

1) include system design alternatives that are incrementally scalable as system throughput increases over the facility lifetime. To be incrementally scalable, system designs shall consider installation of multiple units of a system component (e.g., drive, motor, pump, aeration facility blower) such that: (a) when operated together, all component units can be operated in an energy efficient manner at the maximum design loading specifications and (b) when operated incrementally at loadings below the design specification, the component units can **also** be operated in a way to maximize energy efficiency.
2) identify three project alternatives with high net present value (i.e. considering capital and operating costs) based on life cycle cost analyses (LCCA) and for each calculate the net present value of energy savings. The energy savings attributable to each process component (e.g., drive, motor, and method of control) shall be evaluated separately and together as a whole system (all components integrated). The LCCA shall evaluate the energy savings attributable to incrementally scalable system design alternatives defined in (1), above, and shall include a sensitivity analysis that considers how energy supply cost volatility may affect the LCCA.

Participant Discussion:

Discussion Topic 1: Comments regarding the need for design guideline energy efficiency wording. Would such wording support regional, state, and local initiatives to advance energy efficiency?

Discussion Topic 1 Summary: Currently, participants are implementing energy efficiency options as a standard practice in wastewater treatment system design. For King County, energy efficiency wording is in the county's Orange Book that governs WWTF design. In Green Bay, Chicago, and San Antonio, energy efficiency considerations are standard practice without the force of energy efficiency wording in design guideline code. Energy efficiency project implementation may be impeded by factors such as 1) oversizing capacity relative to near-term loads so that the wastewater system can handle higher loads over a 20-year plus system life, 2) building redundancy into a wastewater system to ensure reliability, and 3) low bid rules that make it more difficult to implement more expensive energy efficiency alternatives.

Green Bay Metropolitan Sewage District – Dan Busch indicated that the following barriers often impede energy efficiency in WWTF design: 1) life-cycle design requirement that pushes plant designers to oversize system components to enable the system to handle load growth over a 20-year planning horizon and 2) system reliability requirement that pushes plant designers to incorporate redundancy into the system, possibly increasing energy use. Mr. Busch indicated that a municipality could address oversizing and system reliability design hurdles, resulting in energy-efficient facility design, by incorporating multiple smaller system components (e.g., pumps and motor drives).

King County Wastewater Treatment Division – Roger Browne indicated that use of high-efficiency motors is a current design requirement, per Orange Book guidelines. The Orange Book (*Criteria for Sewage Works Design*, http://www.ecy.wa.gov/biblio/9837.html, January

1978 [revised August 2008]) governs WWTF design in Washington. The Orange Book, under the Requirements for Engineering Report and Facility Plan, also states that the facility plan alternatives analysis is subject to the same requirements as the engineering report "plus description of innovative and alternative technologies [that is, those saving energy and nonconventional treatment (land application, etc.)]". King County facilities are designed for 20 to 25 years of operation.

Roger Browne also felt that politicians will not incorporate energy efficiency wording into regional, state, and local design guidelines unless energy efficiency becomes a more important concern to them. Currently, KCWTD effectively has such wording in place, in the form of a goal to increase energy efficiency by 10% by 2012. King County has an energy manager for wastewater. One KCWTD facility is implementing a $12 million methane gas to electricity project. KCWTD also is constructing a Leadership in Energy and Environmental Design (LEED) green building with a passive solar wall.

Metropolitan Water Reclamation District of Greater Chicago – Chris Nam indicated that MWRDGC utility costs have not been rising because of successful efforts to implement capital investments to improve energy efficiency. MWRDGC is constructing a modified LEED facility.

San Antonio Water System – Meg Conner indicated that a low-bid procurement requirement was the primary barrier to implementing energy-efficient design practices in WWTFs. Meg Conner and Dan Titerle indicated that energy-efficient design practice is standard practice for SAWS. For example, use of high-efficiency motors is a current design requirement. SAWS facilities are designed for 20 to 50 years of operation.

Discussion Topic 2: Comments on design guideline energy efficiency wording.

Discussion Topic 2 Summary: Wording comments fall into three categories:

1. When should an evaluation of energy efficiency alternatives be required? Roger Browne (King County) suggests that such evaluations should not apply to "a facility reevaluation for capacity rating".
2. System performance wording. Roger Browne (King County) and Dale Doerr (Sheboygan) stated wording preferences for "loading" and "throughput", which appear in the draft wording. Participants discussed whether an "incrementally scalable" WWTF could be designed such that "(2) when operated incrementally at loadings below the design specification, the component units **also** can be operated in a way to maximize energy efficiency". In his written comments, Dan Busch (Green Bay) emphasized that an incrementally scalable design must address system reliability, redundancy, and maximum energy efficiency, collectively.
3. Economic burden of developing multiple design options. Chris Nam (Chicago) indicated that upfront design costs limit the number of evaluations and that a requirement to evaluate three project alternatives would present an economic burden.

Green Bay Metropolitan Sewage District – Attachment 1 includes Dan Busch's written comments. His suggested wording emphasizes that an incrementally scalable design must address system reliability, redundancy, and maximum energy efficiency.

King County Wastewater Treatment Division – Roger Browne suggested that the first sentence in the draft energy efficiency wording should not include "a facility reevaluation for capacity rating". He indicated that the first sentence needs clarification regarding the timing of an evaluation of energy efficiency alternatives.

Roger Browne had a preference for replacing the word "loadings" with "throughput" in the following draft energy efficiency wording phrase: "(2) when operated incrementally at loadings below the design specification, the component units can **also** be operated in a way to maximize energy efficiency".

Metropolitan Water Reclamation District of Greater Chicago – Chris Nam indicated that upfront design costs limit the number of design alternatives they can develop. Design guideline wording that specified evaluating "three project alternatives" would be a cost burden.

Sheboygan Regional Wastewater Treatment Facility – Attachment 1 includes Dale Doerr's written comments. He prefers that "loading" be used instead of "throughput" in the following draft energy efficiency design guideline wording: "(1) include system design alternatives that are incrementally scalable as system throughput increases over the facility lifetime".

Can a WWTF be designed to be energy efficient at all flow rates? Participants discussed whether it was possible to design a system such that, per the following draft energy efficiency design guideline wording excerpt, "(2) when operated incrementally at loadings below the design specification, the component units can **also** be operated in a way to maximize energy efficiency".

Roger Browne (KCWTD): Because of variation in wastewater flow rate, pumps in the wastewater treatment system often are not operating at the most energy-efficient part of their load curves. Variable frequency drives cannot keep a pump operating at the most energy-efficient part of the load curve as flow rates vary.

Chris Nam (MWRDGC): A facility can optimize energy efficiency in pumping systems by sizing the pumps to operate at the most efficient part of their load curves over the most likely flow range.

Meg Conner (SAWS): Ms. Conner agreed that a facility can optimize energy efficiency for a pumping system for a likely load range. Energy efficiency will fluctuate with load variation.

Joe Cantwell (SAIC): A facility can optimize pumping system energy efficiency by addressing motor scalability. Rather than designing the pumping system with fewer oversized motors, design the system with more motors that can be placed online incrementally as load increases.

Bob Lorand (SAIC): A facility can optimize pumping system energy efficiency by making it an initial design concern.

Discussion Topic 3: Pathways to incorporating energy efficiency wording into regional, state, and local design guidelines, plus the barriers to incorporating wording and approaches to addressing those barriers.

Discussion Topic 3 Summary: The participants discussed the SRF as a pathway to implementing energy efficiency projects. Specifically, energy efficiency project implementation could be assisted by 1) including energy efficiency as an SRF loan evaluation criterion, 2) offering lower SRF loan interest rates for energy efficiency projects, and 3) lowering the principal to be repaid on an energy efficiency project loan through principal forgiveness authorized by the SRF. The SRF pathway is a possible alternative to a pathway for adopting design guideline energy efficiency wording.

State Revolving Fund Loan Application and Loan Practices to Support Energy Efficiency

Meg Conner (SAWS): The SRF could promote energy-efficient design in WWTFs by including energy efficiency as an evaluation criterion in construction loan applications.

Dale Doerr (SRWTF): In some municipalities, such as Sheboygan, it is uncertain whether the SRF would approve a construction project justified solely on improvement of energy efficiency.

Dan Busch (GBMSD): The SRF could promote energy-efficient design in WWTFs by 1) including energy efficiency as an evaluation criterion in construction loan applications and 2) offering reduced interest rates for energy efficiency projects. The SRF would be likely to approve a construction project justified as system expansion even though the project's actual intent was to implement capital investments that boosted system energy efficiency.

Joe Cantwell (SAIC): Principal forgiveness is another option that the SRF could use to lower the cost of capital investments to improve system energy efficiency.

Cautionary Note Regarding Efforts to Increase Energy Efficiency in Wastewater Treatment Systems

Meg Conner (SAWS): Evolving regulations may require that wastewater treatment systems move to more energy-intensive processes, thus diminishing system energy efficiency. For example, regulations are pushing treatment systems away from chlorination and toward a more energy-intensive UV process.

Attachment 1: Energy Efficiency Wording Editorial Comments:

Following are written comments regarding the draft design guideline energy efficiency wording. SAIC sent a request for comments to 10 organizations, primarily municipal wastewater treatment organizations. Comments include responses to the following questions:

1. How might such wording affect your organization's wastewater treatment facility design effort?
2. How might such wording affect management of your organization's wastewater treatment facilities?
3. Would such wording support regional/state/local initiatives to advance energy efficiency?
4. What pathways might be used to incorporate energy efficiency language into regional/state/local design guidelines?
5. What are the issues or barriers that need to be addressed prior to incorporating energy efficiency wording into design guidelines, and what are your suggested approaches to addressing these barriers?

Green Bay Metropolitan Sewage District – Comments from Dan Busch

"I do believe this is a very good initiative and much needed; however, it will have to overcome what I call 'Conservative Engineering', i.e., engineers never design anything to fail and the 'safety margin' and 'risk reductions' that factor into design often override the recognition of the need to 'conserve energy'". (Source: e-mail from Dan Busch [Director of Operations, Green Bay Metropolitan Sewage District] to Joseph C. Cantwell, William R. King, and Robert T. Lorand, July 29, 2009)

Mr. Busch, in a memorandum, states: "I feel wording suggested below should be mandatory for Green projects, energy efficiency projects and any government supported grant programs that advocate energy utilization and incorporation of renewable energy. You may want to advocate that States provide an interest deduction under the State Revolving Loan Fund (SRLF) for projects that meet energy design standards. My feeling is many municipalities are not able to recognize the value of annual account energy savings impact on the project payback over 20 years".

Draft Wording: All projects requiring a facility plan, facility plan amendment, comparison of alternatives, facility modification, or a facility reevaluation for capacity rating shall:

1) include system design alternatives that are incrementally scalable as system throughput increases over the facility lifetime. To be incrementally scalable, system designs shall consider installation of multiple units of a system component (e.g., drive, motor, pump, aeration facility blower) such that: (a) when operated together, all component units can be operated in an energy efficient manner at the maximum design loading specifications and (b) when operated incrementally at loadings below the design specification, the component units can **also** be operated in a way to maximize energy efficiency. **The incrementally scalable design shall recognize and support the requirements for system reliability and redundancy and demonstrate the collaborative relationship between reliability, redundancy, and maximum energy efficiency. These three factors should support each other, not compete with each other.**

2) identify three project alternatives with high net present value (i.e. considering capital and operating costs) based on life cycle cost analyses (LCCA) and for each calculate the net present value of energy savings. The energy savings attributable to each process component (e.g., drive, motor, and method of control) shall be evaluated **both** separately and together as a whole system (all components integrated). The LCCA shall evaluate the energy savings attributable to incrementally scalable system design alternatives defined in (1), above, and shall include a sensitivity analysis that considers how energy supply cost volatility may affect the LCCA.

In that memorandum, in response to a request for written comments on the above draft energy efficiency and management guidance wording and in answer to several related questions, Mr. Busch replied as follows:

1. How might such wording affect your organization's wastewater treatment facility design effort?

   "Making it a 'need to have' in a facility plan rather than a 'nice to have' would eliminate the internal debate as to when energy efficiency should be an attribute when selecting design alternatives or equipment".

2. How might such wording affect management of your organization's wastewater treatment facilities?

   "In parallel with the answer above, such wording would put everyone on the same page and make it mandatory that energy efficiency be an evaluating criterion when selecting equipment and designing processes".

3. Would such wording support regional/state/local initiatives to advance energy efficiency?

   "As stated above, if they tie energy efficiency into reduced loan rates under the SRLF, municipalities would have an upfront incentive to consider meeting some level of energy efficiency standards".

4. What pathways might be used to incorporate energy efficiency language into regional/state/local design guidelines?

   "Make it incentive-based right up front and not have the municipality take the risk that the calculated energy savings are going to materialize. There will need to be a mechanism to ensure that systems designed to be energy efficient are operated as such, but my general experience is that people will operate their plants to save energy. They just need the incentive to get their governing boards to buy into the process".

5. What are the issues or barriers that need to be addressed prior to incorporating energy efficiency wording into design guidelines, and what are your suggested approaches to addressing these barriers?

"Issues are that many times the payback to incorporate energy efficiency may not meet internal standards that are used to set what is an acceptable payback period.

Evaluating O&M costs cannot be based on assuming static energy prices; need to have a mechanism that accounts for escalating energy costs over the period of the LCCA.

Get design engineers more comfortable with incorporating energy efficiency and incremental systems into designs. My experience is that the conservative nature of designing and applying safety factors makes it difficult for scalable systems to be implemented and meet energy efficiency criteria".

Source: Memorandum from Dan Busch (Director of Operations, GBMSD, to Joseph C. Cantwell, William R. King, and Robert T. Lorand, July 30, 2009)

King County Wastewater Treatment Division (Seattle, Washington)

Following are King County's comments on the draft design guideline energy efficiency wording and responses to survey questions:

Draft Wording: *All projects requiring a facility plan, facility plan amendment, comparison of alternatives, facility modification, or a facility reevaluation for capacity rating shall:*
[In terms of energy efficiency, projects or modifications that do not involve equipment or processes that require energy would be unsuitable for the proposed analysis.]
"All projects that construct new treatment works or pump/lift station, or that modify the equipment or capacity of existing treatment works or pump/lift station shall:"
1) include system design alternatives that are incrementally scalable as system throughput increases over the facility lifetime. To be incrementally scalable, system designs shall consider installation of multiple units of a system component (e.g., drive, motor, pump, aeration facility blower) such that: (a) when operated together, all component units can be operated in an energy efficient manner at the maximum design loading specifications and (b) when operated incrementally at loadings below the design specification, the component units can ***also*** *be operated in a way to maximize energy efficiency.*
["Maximum design loading" is not likely the operating point that will produce the best energy savings over the life of a facility. Perhaps an analysis can be based on a "flow-weighted daily average". An analysis also would require specifying hydraulic loading, mass loading, or some combination thereof; and the analysis may need to specify the increment of scalability.]
"(1) ...consider installation of multiple units of a system component...such that all component units can be operated in a way to maximize energy efficiency at flow levels of both maximum design loading specifications and below maximum design loading specifications".
Possibly more beneficial might be to say "system component units shall be incorporated so as to maximize energy efficiency at modal flow rates, but also take into account energy efficiency at maximum design loading specifications".
1) include system design alternatives that are scalable as system throughput increases such that when operated together, component units and major subsystems can be operated in an energy efficient manner over the expected lifetime of the facility.
2) identify three project alternatives with high net present value (i.e. considering capital and operating costs) based on life cycle cost analyses (LCCA) and for each calculate the net present value of energy savings. The energy savings attributable to each process component (e.g., drive, motor, and method of control) shall be evaluated separately and together as a whole system (all components integrated). The LCCA shall evaluate the energy savings attributable to incrementally scalable system design alternatives defined in (1), above, and shall include a sensitivity analysis that considers how energy supply cost volatility may affect the LCCA.
[No particular reason is apparent for specifying exactly three alternatives, but it seems impractical to limit in either direction. All alternatives assessed should be subjected to LCCA.]
Instead of focusing on NPV, why not focus on LCCA, which already accounts for lower energy costs or O&M costs associated with energy savings? When implementing new energy efficiency projects, a lower LCCA illustrates success just as effectively as NPV.
Along the lines of LCCA or NPV, sometimes a very energy-efficient alternative is available, but it might not meet process and permit compliance requirements. Municipalities need to consider situations such as these in the selection of alternatives, as well.]
(2) all project alternatives shall be subjected to a life cycle cost analysis (LCCA) that considers capital and operating cost in addition to functional requirements such as process efficiency, reliability and permit compliance. The energy usage attributable to each process component (e.g., drive, motor, and method of control) shall be evaluated separately and together as a system with all of its components integrated. The LCCA shall evaluate the energy usage attributable to scaled system design alternatives defined in (1), above, and shall include a sensitivity analysis that considers how energy supply cost volatility may affect the LCCA. Energy to run treatment works is a precious resource. Where other project considerations are equal, give preference to lower energy usage alternatives and projects that recover resources.

Responses to Questions:

1. How might such wording affect your organization's wastewater treatment facility design effort?

   Because we already have a significant effort focused on green building and operations and a great deal of internal guidance on the topic, this wording likely would not affect our own facility design effort. However, this wording might have significant impact on smaller sewage treatment operations. As Ms. Schweitzer identified in the meeting, reaching out to a wide cross-section of such operating entities would likely be very helpful for organizations that do not already have efforts in motion.

2. How might such wording affect management of your organization's wastewater treatment facilities?

   King County is already committed to energy-efficient operation, resource recovery, LEED considerations for new buildings, renewable energy, and emerging technology in these areas. Because it is similar to the first response, the proposed wording would not likely change the management of our wastewater operations. We already are focusing on optimizing them in terms of energy efficiency, but we imagine that it could help other entities greatly.

3. Would such wording support regional/state/local initiatives to advance energy efficiency?

   Yes, such language would complement current efforts.

4. What pathways might be used to incorporate energy efficiency language into regional/state/local design guidelines?

   We recommend that efforts begin with the Washington State Department of Ecology or its equivalent in each state. The department will have contacts at lower levels, and it tends to issue policy and guidance documents that would advance this initiative.

5. What are the issues or barriers that need to be addressed prior to incorporating energy efficiency wording into design guidelines, and what are your suggested approaches to addressing these barriers?

   This is an interpretation and translation exercise. Consider creating specification language samples, which would be an area that would help many resource-strapped entities that do not necessarily have the time or expertise to focus on incorporating energy efficiency. Many operations across the United States lack the resources to do much more than operate what they have. They also contract design work to engineering consultants. Approach professional engineering organizations for input and for assistance in spreading the word. Consider a process equipment or system rating equivalent to the Energy Star building rating (a brand that would signify WERF evaluation or endorsement). However, because efficiency equipment by itself does not make an efficient system, it may be most beneficial to have a rating for the complete system that is similar to Energy Star buildings ratings.

Source: Suzanne Schweitzer, Roger Browne, Carl Grodnik, and Bruce Kessler, "Comments from King County Washington", included in e-mail from Suzanne Schweitzer, King County, to William R. King, SAIC, August 25, 2009. The file presents the composite comments of members of the King County Wastewater Division: Suzanne Schweitzer (Resource Recovery Section Manager, Wastewater Treatment Division), Roger Browne (Wastewater Treatment Division), Carl Grodnik (Wastewater Treatment Division's energy manager), Bruce Kessler (Wastewater Treatment Division Unit Manager for Engineering and Technical Resources)

Sheboygan Regional Wastewater Treatment Facility – Comments from DALE DOERR

> Draft Wording: All projects requiring a facility plan, facility plan amendment, comparison of alternatives, facility modification, or a facility reevaluation for capacity rating shall:
> 1) include system design alternatives that are incrementally scalable as system throughput (loading) increases over the facility lifetime. To be incrementally scalable, system designs shall consider installation of multiple units of a system component (e.g., drive, motor, pump, aeration facility blower) such that: (a) when operated together, all component units can be operated in an energy efficient manner at the maximum design loading specifications and (b) when operated incrementally at loadings below the design specification, the component units can also be operated in a way to maximize energy efficiency.
> 2) identify three project alternatives with high net present value (i.e. considering capital and operating costs) based on life cycle cost analyses (LCCA) and for each calculate the net present value of energy savings. The energy savings attributable to each process component (e.g., drive, motor, and method of control) shall be evaluated separately and together as a whole system (all components integrated). The LCCA shall evaluate the energy savings attributable to incrementally scalable system design alternatives defined in (1), above, and shall include a sensitivity analysis that considers how energy supply cost volatility may affect the LCCA.

1. How might such wording affect your organization's wastewater treatment facility design effort?

   We look at energy efficiency in every aspect of WWTP operations, including plant design. The impact on our operations would be minimal.

2. How might such wording affect management of your organization's wastewater treatment facilities?

   Same as above; it would not affect management of our WWTP.

3. Would such wording support regional/state/local initiatives to advance energy efficiency?

   Yes, some facilities will not make the extra effort toward energy efficiency, unless they are forced to do so by design criteria.

4. What pathways might be used to incorporate energy efficiency language into regional/state/local design guidelines?

   Making energy efficiency a requirement of facility plan approval is a start. Adding this requirement as mandatory in Wisconsin Department of Natural Resources (WDNR) design criteria would provide additional incentive for design engineers and owners (municipal entities) to live energy efficiency.

5. What are the issues or barriers that need to be addressed prior to incorporating energy efficiency wording into design guidelines, and what are your suggested approaches to addressing these barriers?

   The number one barrier will be the added up-front costs associated with installing energy-efficient equipment. Owners (city officials) must understand that many municipal WWTP operations are monopolies and, unlike the private sector, the need for a short payback period is not necessary. Energy efficiency has both an immediate and long-term payback.

APPENDIX C

OUTLINE FOR ENERGY BEST PRACTICE GUIDEBOOK

The following sections relevant to wastewater are from *Water and Wastewater Industry Energy Best Practice Guidebook,* Focus on Energy, prepared by Science Applications International Corporation, http://www.focusonenergy.com/Business/Industrial-Business/Guidebooks/default.aspx , December 2006. The WERF website link to this guidebook is:

http://www.werf.org/AM/TemplateRedirect.cfm?template=/CM/ContentDisplay.cfm&ContentID=10245

The last appendix entry in the following list, Appendix W, references WERF software. While not included in the Focus on Energy guidebook, it will be added if a revised version of the guidebook is developed under WERF funding.

INTRODUCTION

Energy Use in Wastewater Treatment and Collection Systems

ENERGY USE

Energy Baseline

Energy Benchmark

MANAGEMENT BEST PRACTICES

TECHNICAL BEST PRACTICES

Wastewater Treatment Best Practices
- Variable Frequency Drive Applications
- Reduce Fresh Water Consumption
- Optimize Flow with Controls
- Operational Flexibility
- Staging of Treatment Capacity
- Manage for Seasonal/Tourist Peaks
- Flexible Sequencing of Tank Use
- Recover Excess Heat from Wastewater
- Cover Basins for Heat Retention
- Optimize Aeration System
- Fine-Bubble Aeration
- Aerobic Digestion Options
- Biosolids Processing Options
- Biosolids Mixing Options: Aerobic
- Variable Blower Air Flow Rate: Aerobic
- Dissolved Oxygen Control: Aerobic
- Biosolids Mixing Options: Anaerobic
- Ultraviolet (UV) Disinfection Options
- Final Effluent Recycling

General Facility Best Practices

Facility Energy Assessments
Real-Time Energy Monitoring
Energy Education for Facility Personnel
Electric Peak Reduction
Manage Electric Rate Structure
Idle or Turn Off Equipment
Install High-Efficiency Motors
Variable Speed Technologies
Optimize Pump System Efficiency
Comprehensive Planning Before Design
Design Flexibility for Today and Tomorrow
Renewable Energy Options

APPENDICES

Appendix C: Wastewater Facilities from Low Efficiency to High Energy

Appendix D: Best Practices for Common Systems

Appendix E: Examples of Energy Saving Utilities in Wisconsin

Appendix F: Variable Speed Technology Options

Appendix I: Additional Resources

Appendix W: WERF Available Software

Life-Cycle Assessment Manager for Energy Recovery (LCAMER)

Green Energy Life-Cycle Assessment Tool (GELCAT)

APPENDIX D

WASTEWATER TREATMENT FACILITY ENERGY REDUCTION PROGRAM SUMMARIES

Alliance To Save Energy: Watergy™ Program[6]

Website: http://www.watergy.org/

Contact: Laura Van Wie McGrory, Director

E-mail: lvanwie@ase.org

Program Overview: Watergy projects enable water supply systems to achieve energy and water savings through implementation of technical and managerial solutions.Watergy projects, which involve the delivery of quality water, are spread worldwide in developing countries and have been implemented in more than 40 cities.

Key Concepts: Help municipal authorities understand the linkage between water and energy consumption. For example, a reduction in water losses from theft, leakage, or wasteful consumer usage habits will result in lower energy consumption. Help municipal authorities understand that cost-effective measures are available to reduce energy consumption and save water.

Key Projects for Realizing Savings: Improve pumping system; manage leaks; automate system operations; and monitor regularly, preferably with end-use meters.

Payback Range: Less than 1 to 3 years (see http://www.watergy.net/overview/what.php for project examples falling in this range).

Program Implementation Lessons Learned: Successful Watergy projects have benefitted from the following:

- Political will – Senior management for the water supply system not only must be willing to institutionalize improvements, but also must do so to ensure these improvements become standard practice.
- Community outreach – Educate the public to build support and mitigate public opposition.
- Cost-benefit analysis – Use a reputable organization to perform the analysis. The key concept is that the analysis must present economic and technical operating assumptions and payback periods that are acceptable to the local authorities.

---

[6] Watergy, various information from website (http://www.watergy.org/), February 2009.

California Energy Commission: Energy Partnership Program

Website: http://www.energy.ca.gov/efficiency/partnership/index.html

Contact: Shahid Chaudhry, Program Manager, Water-Energy Efficiency

E-mail: Schaudhr@energy.state.ca.us

Program Overview: The California Energy Commission provides energy audits to publically owned water and wastewater facilities to identify and recommend energy-saving measures. The commission provides technical assistance services (up to $20,000 of consultants' costs), usually at no cost to the utilities. The cost of a study depends on facility size, facility type, and scope of the project. The commission contracts with experienced engineering and architectural consultants who provide this technical assistance.

The commission provides technical services to both existing and new construction facilities. Areas in which the program offers services include the following:

Existing Facilities

- Conducting energy audits and preparing feasibility studies
- Reviewing existing proposals and designs
- Developing equipment performance specifications
- Reviewing equipment bid specifications
- Assisting with contractor selection
- Assisting with commissioning

New Construction

- Providing design consultation
- Comparing different technologies
- Reviewing schematics and construction plans
- Providing equipment specification consultation
- Identifying cost-effective, energy-saving measures
- Developing computer simulation models for the planned project
- Helping select experienced professionals with energy efficiency expertise
- Assisting with commissioning.

Key Concepts: Efficiency upgrades reduce annual maintenance costs, increase cost savings, conserve resources, and reduce green house gas emissions.

Energy Partnership Program Participation Cost: Usually none.

Application Process: A very simple, no-hassle application process.

California Energy Commission: Energy Financing Program

Website: http://www.energy.ca.gov/efficiency/financing/index.html

Contact: Shahid Chaudhry, Program Manager, Water-Energy Efficiency

E-mail: Schaudhr@energy.state.ca.us

***Program Overview*:** The California Energy Commission provides low-interest loans to implement energy efficiency measures at publically owned water and wastewater facilities. The implemented projects help utilities to reduce their energy consumption, saving money and reducing greenhouse gas emissions. Projects with proven energy or capacity savings are eligible, provided they meet the eligibility requirements.

A variety of projects are eligible for the loan; examples of projects include:

- Technologies: lighting systems; pumps and motors; streetlights and light emitting diode (LED) traffic signals; automated energy management systems and controls; building insulation; energy generation, including renewable and CHP projects; heating and air conditioning modification; and waste water treatment equipment
- Building or facility type: existing buildings, some new construction, or other energy-using facilities.

Interest Rate: Interest rates for the loans vary. As of December 2009, two interest rates were available: 1% and 3%.

Loan Amount: Loans can finance up to 100% of the project cost, with a maximum loan amount of $3 million.

Loan Criteria: Energy efficiency projects must be technically and economically feasible. Borrowers must repay loans from savings within 15 years, including principal and interest.

Loan Security Requirements: A promissory note, a loan agreement between the utility and the California Energy Commission, and a Tax Certificate are all that is required to secure the loan.

Application Process: A very simple, no-hassle application process.

California Public Utility Commission: California Wastewater Process Optimization Program

Website: http://www.calwastewater.com/index.html

Contact: California Wastewater Process Optimization Program (CalPOP)

E-mail: EfficiencyPrograms@quest-world.com

Address: Quantum Energy Services & Technologies, Inc. (QuEST), 2001 Addison Street, Suite 300, Berkeley, CA 94704

Program Overview:[7] The California Public Utility Commission (CPUC) uses system benefit charge money to fund the CalPOP, which helps to optimize the processes of small (up to 15 MGD[8]) municipal or private WWTFs that are customers of qualified California investor-owned utilities. Investor-owned utilities manage CalPOP through selected contractors. Since 2006, CalPOP has been active only in the PG&E service territory. PG&E selected QuEST to be the program's authorized service provider. The current QuEST contract period of performance is 2009 through 2011. QuEST works with plant operators to perform free process audits, install new equipment, and provide training.

Program Inception: 2000. Currently, the program is in its fifth cycle (2009 through 2011).

Funding Level:[9] $3.7 million for 2006 through 2008; $720,000 to date authorized by the CPUC for 2009 through 2011.

Key Concept: Use system benefit charge funds to provide no-cost and no-obligation facility audits and recommendation reports to facilities.

CalPOP can provide up to 100% of project costs based on energy savings, along with technical assistance, if the facility decides to implement the recommended measures. Projects with greater potential energy savings receive more funding.

CalPOP works with facility staff to implement new processes, install equipment, and provide any needed training.

Key Projects for Realizing Savings: Find opportunities to reduce aeration and pumping costs. Use available funding to upgrade or install new equipment, including dissolved oxygen probes, supervisory control and data acquisition (SCADA) systems, timers, variable frequency drives (VFDs), and other automated controls that tailor the right amount of aeration for plant loading. Provide funding for new energy-saving devices, such as low-powered mixers and aerators, to replace older, less-efficient models.

Payback Range: on average, less than a year

Program Implementation Lessons Learned:[10]

---

[7] E-mail correspondence between William R. King (SAIC) and Shahid Chaudhry (CEC), February 19, 2009.

[8] SBW Consulting, Inc., Impact and Process Evaluation Final Report for QuEST's 2004-5 California Process Optimization Process Program, submitted to CPUC, CALMAC Study ID QST0002.01, May 14, 2007, p. 9.

[9] E-mail correspondence between William R. King (SAIC) and Shahid Chaudhry (CEC), February 19, 2009.

[10] John Bidwell (QuEST), Jennifer Fagan (Itron), Todd Amundson (BPA), "Water-Wastewater Program Best Practices: Lessons from the Field", presented at ACEEE 2008 Summer Study on Energy Efficiency in Buildings, Pacific Grove, CA, August 2008, p. 9.

- Use performance contract–based incentives to attract customers with improved measure cost-effectiveness – pursue cost-effective energy efficiency measures. Program incentives are performance-based; greater kilowatts per hour (kWh) savings yield a higher incentive payment and improved project payback.
- Use customer commitment forms to ensure project completion – for CalPOP, customers that have signed an incentive application and agreement have completed installation. The incentive application and agreement contains legal terms and conditions for completing a project and commits the customer to proceed with the project. The incentive application is part of the audit report package. Results of the no-cost facility audit are contained in the audit report. The no-cost facility audit is performed once the customer signs a site access agreement.
- Use current energy efficiency market drivers relevant to wastewater systems to expand program participation – municipalities are adopting climate change policies. Market wastewater system energy reduction projects are a way to reduce the system's carbon footprint.
- Market the program to both existing and new wastewater system operators to expand program participation – CalPOP has been restricted to process optimization and equipment replacement in existing wastewater systems. A broader market would include existing plant expansions and new construction.

The following tables summarize CalPOP best practices for program design and set-up (Table D-1) and program implementation (Table D-2).

Table D-1. CalPOP Program Design and Set-Up Best Practices.

Program Element	Best Practices
Implementation Approach	• Use a performance-based contract with the implementation contractor. This provides Implementer with a strong incentive to pursue projects with viable candidates.
Program Management	• Program Manager has working knowledge of wastewater operations; focuses on program administration, marketing, and project coordination. • Provide customer with a wastewater engineering consultant who is responsible for identifying measures, preparing audit and providing project management assistance.
Contractor Selection	• Provide customer with a wastewater engineering consultant; use of customer's preferred engineering contractor is potentially problematic. • Identify and maintain a cadre of well-qualified wastewater engineering contractors to recruit customers, conduct analyses, and provide technical assistance and project management.
Program Duration	• Establish a program cycle of three or more years to allow for long project lead times; allow marketing and implementation to bridge between program cycles.
Program Process	• Establish a simplified, stepwise program process with flexibility to address a wide range of customer circumstances. Use simplified program forms to track milestones and solidify customer commitment to project.
Target Markets and Technologies	• Serve a broader market segment. Include new and existing water supply, wastewater and water reclamation facilities, but only if the program team includes appropriate wastewater engineering talent. • Allow for a broad set of measures spanning equipment replacement, process optimization, and plant or capacity expansions for existing sites. Also, include newer on-site generation technologies including fuel cells, cogeneration driven anaerobic digesters using recovered fats, oils, and grease.

Source: John Bidwell (QuEST), Jennifer Fagan (Itron), Todd Amundson (BPA), "Water-Wastewater Program Best Practices: Lessons from the Field", presented at ACEEE 2008 Summer Study on Energy Efficiency in Buildings,

Pacific Grove, CA, August 2008, modified version of Table 3, including program implementation best practices sections from Table 5.

Table D-2. CalPOP Program Implementation Best Practices.

Program Element	Best Practices
Marketing and Outreach	• Perform outreach via industry conferences and trade association meetings • Demonstrate the measure's viability via case studies, preferably those that are local or regional and feature emerging technologies • While some lead generation was sought through attendance and presentations at industry conferences and meetings, workshops considered a luxury within a performance contract • In serving municipal customers, efforts should be made to market to and support the emerging proliferation of greenhouse gas reduction strategies. Private customers are also increasingly interested in documenting emission reductions
Project Identification and Lead Generation	• Rely on the selected market actors who understand program process and screening for lead generation. These include: designated consultant teams, equipment vendors, utility account managers and other trade allies
Project Screening	• Vet candidates by performing site visits and collecting intelligence from wastewater consultant teams, equipment vendors, utility account managers, and other trade allies • Carefully screen candidates, to identify organizations that are proactive, therefore more likely to implement projects • Focus of a wastewater system audit can range from a full menu of measures to only measures in which the customer is interested. Under a performance contract, attempts should be made to limit measures only to what the customer is likely to implement
Project Facilitation	• Provide continuous follow-up throughout project marketing, planning, and implementation • Use a coordinated approach to follow-up with the candidate, working in tandem with assigned wastewater engineering consultant • Target marketing and planning efforts toward operations staff, and later to high-level municipal and public finance authorities during decision phase; return to provide technical assistance to operations staff during project implementation • Focus should always be on addressing customer's barriers to implementation; provide no-cost facility audit, followed by ongoing project technical assistance following commitment
Project Funding and Viability	• Provide access to independent funding sources for organizations that do not have sufficient capital to fund projects • Vet measures that solve operational problems in addition to providing energy benefits. Emphasize non-energy benefits as part of the project justification

Source: John Bidwell (QuEST), Jennifer Fagan (Itron), Todd Amundson (BPA), "Water-Wastewater Program Best Practices: Lessons from the Field", presented at ACEEE 2008 Summer Study on Energy Efficiency in Buildings, Pacific Grove, CA, August 2008, modified version of Table 4, including program implementation best practices sections from Table 5.

Massachusetts Department of Environmental Protection: Energy Management Pilot for Wastewater and Drinking Water Plants

Website: http://www.mass.gov/dep/water/wastewater/empilot.htm

Contact: Michael DiBara, Project Manager

E-mail: Michael.dibara@state.ma.us

Address: MassDEP, 627 Main Street, Worcester, MA 01608

Program Overview: MassDEP will coordinate a targeted pilot project involving a total of 14 facilities, including seven WWTFs and seven drinking water treatment facilities (small, medium, and large) through energy auditing, assessments for renewable and clean energy possibilities (e.g. solar, wind, out-fall turbines, and CHP), and support for implementation of energy-related projects. MassDEP will mobilize an array of public and private partners to provide technical and financial support.

Program Inception: December 2007

Key Concepts: The proposed pilot project has three main goals:

Develop energy management plans for each of 14 pilot facilities that will seek to ***reduce the annual electric consumption and costs of treating waste water and drinking water by 20%.***

Develop one or more models for the strategic use of public funding assistance to significantly reduce or fully finance the capital municipal costs needed for project implementation.

Assess the feasibility of expanding this energy-savings project as a replicable model for other wastewater and drinking water treatment facilities in Massachusetts and elsewhere.

Pilot Project Components:

- **MassDEP Project Oversight and Coordination**: MassDEP will provide project oversight and coordination of the multiple partners.
- **Energy Use Benchmarking**: U.S. EPA Region I conducts initial energy benchmarking at all participating facilities using its new Energy Star benchmarking tool.
- **Energy Audits:** For this pilot project, electric and gas utilities will provide facility energy audits (costing approximately $10,000 each) that will help quantify energy uses and costs and identify potential reductions and savings through conservation. **The utilities have agreed to pay 100% of the cost of the audits, provided their projects pursue funding for low-cost and low-payback energy conservation measures.**
- **Renewable Energy Feasibility Assessments**: The Massachusetts Technology Collaborative 's (MTC) Renewable Energy Trust would provide renewable energy technical assessments to identify any fatal flaws in pursuing wind power, bioenergy, solar, outfall microturbines, and so forth.
- **CHP Analyses:** University of Massachusetts Amherst's Northeast CHP program will screen five municipal wastewater sites for CHP and energy-saving opportunities; it will perform further detailed feasibility analysis on one or two wastewater plants.
- **Financial Support for Implementation**: A variety of Massachusetts government sources of funds could be coordinated to support implementation of conservation and renewable energy projects at these facilities, including MTC Renewable Energy Trust

grants, SRF low-interest loans, Mass Department of Energy Resources' (DOER) Energy Conservation Improvement and Alternative Energy, and possible Environmental Bond Fund grants for debt service.

Payback Range: For many conservation and efficiency projects, as low as two to three years.

New York Power Authority: Anaerobic Digester Gas-Fueled Fuel Cells Projects

Website: http://www.nypa.gov/services/fuel%20cell%20projects.htm

Program Overview: The New York Power Authority (NYPA) funds innovative technology demonstrations that address electricity generation and transmission. Included among its demonstrations are five sites with fuel cells using hydrogen from reformed anaerobic digester gas (Table D-3).

Table D-3. NYPA New Technology Demonstrations (including Emissions Offset Program Demonstrations) – Fuel Cells Using Anaerobic Digester Gas.

Demo Location	Number of 200 kW Fuel Cell Units	Output (kW)	Operating Mode
Yonkers Wastewater Treatment Plant	1	200	Grid-parallel
Wastewater Treatment Plant, Flushing Ave., Brooklyn	2	400	Grid-parallel
Wastewater Treatment Plant, Flatlands Ave., Brooklyn	2	400	Grid-parallel
Wastewater Treatment Plant, Hunts Point, Bronx	3	600	Grid-parallel
Wastewater Treatment Plant, Oakwood Beach, Staten Island	1	200	Grid-parallel

New York State Energy Research and Development Authority: Water and Wastewater Program

Website: http://www.nyserda.org/Programs/Environment/muniwaterwwt.asp#Mission

Website for New York Energy $mart^{SM} Program Evaluation Reports:

http://www.nyserda.org/Energy_Information/evaluation.asp

Contact: Kathleen M. O'Connor, PE, Project Manager, Environmental Research

E-mail: kmo@nyserda.org

Mission: Enable water and wastewater treatment facilities in New York State (NYS) to demonstrate and implement energy-efficient technology solutions to reduce energy use and energy cost.

***Program Inception**:* Since 1980, the program has made available cost-shared funding to demonstrate and develop innovative technologies for the wastewater sector. It made funds available to the water sector beginning in the mid-1990s. Also since the mid-1990s, technical assistance, prescriptive and custom rebates, and performance contracting incentives have been available to water and wastewater systems through various New York Energy $mart^{SM} programs. The program grants performance incentives based on energy (kWh) or demand (kilowatt [kW]) savings. In 2007, NYSERDA initiated the Focus on Water and Wastewater Program, directing activities of the Focus program toward increasing utility participation in NYSERDA programs.

Following are examples of demonstrated technologies that reduce wastewater system energy use and provide on-site electricity generation (reducing purchased power):

Selector Contact Stabilization (SCS) Process Demonstration[11] – This process enables a wastewater system to increase wastewater treatment capacity, reduce aeration energy cost, and enhance process stability. The technology, developed by O'Brien & Gere Engineers, addresses activated sludge biological wastewater treatment processes.

Biogas-Fueled Combined Heat and Power Demonstration – The Gloversville-Johnstown WWTF generates electricity for on-site equipment using two 150 kW internal combustion engine-generator sets fired with digester gas recovered from the two-stage anaerobic digestion process.

Program Goals (July 1, 2006 through June 30, 2011):

- Achieve $2 to 3 million per year energy and cost savings[12]
- Upon completion of all projects, achieve savings that should be "nearly 44,500 MWh of electricity savings and 15,067 kW of peak demand reduction. On average, these projects take five to seven years from conception to implementation."[13]

[11] NYSERDA, Full-Scale Demonstration of Selector-Contact Stabilization Process at Town of Rosendale Wastewater Treatment Plant, Ulster County, New York, Final Report 05-07, report prepared by O'Brien & Gere Engineers, Inc. (Syracuse, NY), website: http://www.nyserda.org/publications/Sector-Contact%20Stabilization%20Report%2005-07-web.pdf, August 2005.

[12] New York Energy $mart^{sm} Program, Quarterly Evaluation and Status Report, Quarter Ending September 30, 2008, Final Report, http://www.nyserda.org/publications/3rd%20Quarter%202008%20Report.pdf, November 2008, Section 5.11 (Municipal Water and Wastewater Efficiency), p. 5-16.

[13] ibid., p. 5-17.

"Since July 1, 2006, $4.64 million has been committed under the targeted water and wastewater initiative. An additional $1.34 million has been awarded for municipal water/wastewater projects under the TA Program."[14]

Table D-4. Project and Funding Status through September 30, 2008.[1]

	Number of Projects Approved	Funds Awarded ($ million)	Co-funding ($ million)
RFP 769 Energy Efficiency Improvements at Water & Wastewater Treatment Plants	1	$0.13	$0.05
RFP 601 (Submetering) [2]	2	$1.1	$0.4
Demonstration Projects (569, 786, 857, 935, 1040, 1171)	16	$2.99	$4.09
Technical Assistance [3]	81	$1.34	$1.34
Technology Transfer	1	$0.1	$0.1

[1] Table does not include metrics on the Energy Smart Focus Project Opportunity Notice (PON)

[2] Funded in part under the general Technical Assistance Program

[3] Funded under the general Technical Assistance Program

Source: New York Energy $mart[sm] Program, Quarterly Evaluation and Status Report, Quarter Ending September 30, 2008, Final Report, http://www.nyserda.org/publications/3rd%20Quarter%202008%20Report.pdf, November 2008, Section 5.11 (Municipal Water and Wastewater Efficiency), Table 5-18.

Municipal Water and Wastewater Efficiency Program – Goals and Achievements through September 30, 2008

Activity	Program Goals (July 1, 2006 through June 30, 2011)	% of Goal Achieved
Issue annual solicitation	Select and fund 25 or more projects Provide assistance to a minimum of 25 municipal wastewater and water treatment facilities in New York	24%
Achievements (July 1, 2006 through September 30, 2008) PON 1040 was issued and 17 proposals were received requesting $3.9 million in NYSERDA funding. In total, five projects were developed from the solicitation; two using system benefits charge (SBC) funds. PON 1171 was issued and 12 proposals were received requesting $3.4 million in NYSERDA funding. Four projects are in the contract-development stage; all will be funded with SBC3 monies. The PON has a second round due date in September 2008. 15 proposals were received in the second round of PON 1171, requesting $3.8 million in NYSERDA funding. PON 1040 – The two SBC-funded projects directly affect three facilities in the near term. PON 1171 – The four SBC-funded projects selected from the first round directly affect three facilities in the near term. Projects recommended for funding come from proposals received in response to the annual solicitation. One goal of the solicitation is that all projects selected ultimately produce results with widespread applicability to the municipal wastewater and water sector in New York; this is referred to as "long term assistance". All six projects are anticipated to provide assistance to multiple municipal wastewater and water treatment facilities in New York in the long term.		
Technology transfer	Provide critical information on ways to optimize energy use at municipal wastewater and water treatment facilities.	100%

[14] ibid., p. 5-17.

	Provide information to 1,000 individuals serving the municipal wastewater and water treatment sector in New York.	
Achievements July 2006–December 2006 Four presentations were given throughout the State as part of the NY Co-funding for Water and Sewer Infrastructure conferences. The total attendance for the four conferences was approximately 300 individuals. A presentation was given as part of a webcast hosted by the Comptroller's Office. An energy management training conference was co-developed with Global Energy Partners (an offshoot of EPRI) and the New York Water Environment Association (NYWEA). Approximately 70 individuals (municipal operators and elected officials, consultants, engineers) attended the two-day session held in Cooperstown in November. **Achievements January 2007–December 2007** The submetering and evaluation of 20 wastewater treatment plants were completed. The final site reports and summaries of findings were posted online. Four presentations were given throughout the State as part of the NY Co-funding for Water and Sewer Infrastructure conferences. The total attendance for the four conferences was approximately 300 individuals. An Energy Management issue of *Clearwaters* (published by NYWEA) was developed. NYWEA is the NYS chapter of the nation's premier professional organization for the wastewater treatment profession (Water Environment Federation). The Energy Management issue will be published in spring. **Achievements January 2008–March 2008** Five presentations were given to diverse audiences: Three to Congresswoman Gillebrand's constituency, one at the annual NYWEA conference in NYC and another to local elected officials in White Plains. In total, approximately 300 individuals attended the presentations **Achievements April 2008–June 2008** The Energy Management issue of *Clearwaters* (developed by NYSERDA) was published. Five presentations were given during the second quarter of 2008; at the NY section AWWA spring meeting; one as part of the Genesee/Finger Lakes Regional Planning Commission's summer conference; one as part of the Adirondack Research Consortium annual meeting; one during the NYWEA spring meeting; and one at the national AWWA conference. In total, approximately 250 individuals attended the presentations. **Achievements from June 2008–September 2008** An article is under development for the winter issue of *Clearwaters*. Achievements On-going The Energy Smart Focus program is providing customized services to support energy efficiency in the sector. The program offers outreach materials and training to individuals associated with the sector statewide.		
Energy and cost savings	$2-3 million per year	See Section 5.12.4.
Technical Assistance	Develop, review and approve 30 projects	30%
July 2006–December 2007 Five projects were approved to begin work totaling $112K in NYSERDA funds. Five projects totaling $63K in NYSERDA funds were completed. **January 2008–March 2008** A project totaling approximately $4.5K in NYSERDA funds was completed. **April 2008–June 2008** Two projects were approved to begin work totaling $35K in NYSERDA funds. Two projects totaling approximately $60K in NYSERDA funds were completed. **July2008–September 2008** Two projects were approved to begin work totaling approximately $39K in NYSERDA funds. One project totaling approximately $16K in NYSERDA funds was completed.		

Source: New York Energy $mart[sm] Program, Quarterly Evaluation and Status Report, Quarter Ending September 30, 2008, Final Report, http://www.nyserda.org/publications/3rd%20Quarter%202008%20Report.pdf, November 2008. Section 5.11 (Municipal Water and Wastewater Efficiency), Table 5-17, pp. 5-15 through 5-16.

Municipal Water and Wastewater Efficiency Program – Goals and Achievements through June 30, 2008

<table>
<tr><th>Activity</th><th>Program Goals
(July 1, 2006 through June 30, 2011</th><th>% of Goal Achieved</th></tr>
<tr><td>Issue annual solicitation</td><td>Select and fund 25 or more projects.
Provide assistance to a minimum of 25 municipal wastewater and water treatment facilities.</td><td>24%</td></tr>
<tr><td colspan="3">Achievements (July 1, 2006 through June 30, 2008)
PON 1040 was issued and 17 proposals were received requesting $3.9 million in NYSERDA funding. Five projects were developed from the solicitation; two using SBC funds.
PON 1171 was issued and 12 proposals were received requesting $3.4 million in NYSERDA funding. Four projects are in the contract-development stage; all will be funded with SBC3 monies. The PON has a second round due date in September 2008.
PON 1040 generated two SBC-funded projects directly affecting three facilities in the near term.
PON 1171 generated four SBC-funded projects. While these projects will directly affect facilities in the near term, they were selected also to provide long-term assistance to projects that will produce results with widespread applicability to the municipal wastewater and water sectors in New York. Projects recommended for funding come from proposals received in response to an annual solicitation.</td></tr>
<tr><td>Technology transfer</td><td>Provide critical information on ways to optimize energy use at municipal wastewater and water treatment facilities.
Provide information to 1,000 individuals serving the municipal wastewater and water treatment sector in New York.</td><td>100%</td></tr>
<tr><td colspan="3">Achievements July 2006–December 2006
Four presentations were given to approximately 300 individuals throughout the state as part of New York co-funding for Water and Sewer Infrastructure conferences.
A presentation was given as part of a Webcast hosted by the Comptroller's Office.
An energy management training conference was co-developed with Global Energy Partners, a firm associated with the Electric Power Research Institute, and the New York Water Environment Association (NYWEA). Approximately 70 individuals, including municipal operators, elected officials, consultants, and engineers, attended the two-day session held in Cooperstown in November.
Achievements January 2007–December 2007
The submetering and evaluation of 20 wastewater treatment plants were completed. The final site reports and summaries of findings were posted online.
Four presentations were given to approximately 300 individuals throughout the state as part of New York co-funding for Water and Sewer Infrastructure conferences.
Achievements January 2008–March 2008
Five presentations were given to approximately 300 individuals – three to Congresswoman Gillebrand's constituents, one at the annual NYWEA conference in New York City, and one to local elected officials in White Plains.
Achievements April 2008–June 2008
An Energy Management issue of NYWEA's journal Clearwaters was developed by NYSERDA staff. NYWEA is the new state chapter of the nation's premier professional organization for the wastewater treatment profession, the Water Environment Federation. The Energy Management issue was published in spring 2008.
Five presentations to approximately 250 individuals were given during the second quarter of 2008:
• The New York section of the American Water Works Association (AWWA) spring meeting
• The Genesee/Finger Lakes Regional Planning Commission summer conference
• The Adirondack Research Consortium annual meeting
• The NYWEA spring meeting
• The national AWWA conference</td></tr>
</table>

Ongoing The Energy Smart Focus program (see Section 3.7) provides customized services to support energy efficiency in the water and wastewater sectors. The program offers outreach materials and training to individuals statewide.		
Technical Assistance	Develop, review, and approve 30 projects	23%
Achievements July 2006–December 2007 Five projects receiving $112,000 in NYSERDA funds were approved to begin work. Five projects that received $63,000 were completed. **Achievements January 2008–March 2008** A project using approximately $40,500 in NYSERDA funds was completed. **Achievements April 2008–June 2008** Two projects were approved to begin work using $35,000 in NYSERDA funds. Two projects using approximately $60,000 in NYSERDA funds were completed.		

Source: New York Energy $mart^{sm} Program, Quarterly Evaluation and Status Report, Quarter Ending June 30, 2008, Final Report, http://www.nyserda.org/energy_information/Final%202nd%20Quarter%20Report%208.29.08.pdf, August 2008. Section 5.11 (Municipal Water and Wastewater Efficiency), Table 5-16, pp. 5-14 through 5-15.

Program Design and Set-Up:

Implementation Approach – Solicit wastewater system projects through PON procurements.

Program Management – Place the program under the management of NYSERDA staff with support from competitively selected contractors that provide Outreach and Project Consulting Services and Technical Assistance Services to commercial and industrial customers in NYS.

Contractor Selection – Select Energy $mart^{sm} Outreach and Project Consulting Services contractors and Technical Assistance Services contractors through a competitive bidding (request for proposal [RFP]) process. In July 2007, NYSERDA selected Red Oak Consulting (a division of Malcolm Pirnie, Inc.) as its contractor for developing and implementing the Focus on Municipal Water and Wastewater Facility Energy Efficiency program (http://www.nyserda.org/Programs/Environment/muniwaterwwt.asp#Mission).

Program Implementation:

Marketing and Outreach – The wastewater system operators and project design team need to justify a project to the municipal board responsible for authorizing project funding and to local elected officials. A function of the Water Wastewater Focus contractor is to provide wastewater system operators and the project design team with the tools to communicate project benefits. Such tools may be directed toward audio and visual aids for effective presentation of a project concept, models to demonstrate clearly the project's life-cycle economic benefits, and models to demonstrate alternatives for project financing.[15]

[15] Teleconference, William R. King (SAIC), Matthew Yonkin (Malcolm Pirnie, Inc.), Silvia Marpicati (Malcolm Pirnie, Inc.), discussion of program set-up, design, and implementation lessons learned with NYSERDA Water Wastewater Focus contractor Malcolm Pirnie, Inc., February 2009.

Pacific Gas & Electric: Water Treatment Program

Website:
http://www.pge.com/mybusiness/energysavingsrebates/incentivesbyindustry/wastewater/

Contact: Stephen Fok, Senior Program Engineer

E-mail: skf2@pge.com

Program Overview: PG&E offers WWTFs financial incentives to implement energy-efficient measures through its Non-Residential New Construction (NRNC) and Non-Residential Retrofit (NRR) programs. These programs address new construction, retrofit, and replacement at failure projects.

Program Implementation: The following programs support activities that promote energy efficiency in WWTFs:

- Integrated energy audit of WWTFs – Audits are provided as part of the PG&E Customer Energy Efficiency Program. PG&E uses the results of energy audits as a reference for providing rebates for implementation of energy efficiency, as well as for demand response measures in the audited facilities.
- NRNC – The NRNC program is PG&E's version of the statewide Savings by Design new construction program, which is funded with Public Purpose Programs surcharge money collected by the CPUC.[16] Incentives that apply to wastewater facilities fall under the systems approach part of the program, rather than under the whole building approach part of the program. The program offers free design assistance, covering a range of services from plan review to energy modeling and financial analysis of energy-efficient options. Within the systems approach, financial incentives fall under other systems and processes (processes not covered by Title 24). Under the systems approach/process system category, the 2009 incentives are $0.09/kWh ($1/therm for gas) for annualized energy savings and $100/peak kW for demand reduction, with a $500,000 per project maximum incentive cap (http://www.savingsbydesign.com/overview.htm, September 2009).
- NRR – This PG&E standard performance contract program provides incentives for using energy-efficient equipment in retrofit projects. The same 2009 incentives apply as for the Savings by Design new construction program (i.e., $0.09/kWh [$1/therm for gas] for annualized energy savings and $100/peak kW for demand reduction, with a $500,000 per project maximum incentive cap [http://www.spc-nrrdr.com/download/2009SPCDocs/PGE/ 2009%20PG&E%20app%20forms.xls, September 2009]).
- CalPOP – Since 2006, CalPOP has been active only in the PG&E service territory. PG&E selected QuEST to be the program's authorized service provider. The current QuEST contract period of performance is 2009 through 2011. QuEST works with plant operators to perform free process audits, install new equipment, and provide training. PG&E added an Anaerobic Digester Optimization Pilot (ADOP) program component to its CalPOP activity. This report discusses the CalPOP program separately.

[16] The Public Purpose Programs surcharge money collected by the CPUC funds Savings by Design nonresidential new construction programs operated by PG&E, San Diego Gas and Electric (SDG&E), Southern California Edison (SCE), Southern California Gas Company (SoCalGas), and Sacramento Municipal Utility District (SMUD).

- KEMA Wastewater Process Efficiency Initiative (WPEI) – Conducted for PG&E, this program operated at least from 2006 through 2008.

Key Projects for Realizing Savings: Provide or incorporate dissolved oxygen control, VFDs, premium-efficiency motors, efficient aeration blowers, efficient effluent pumping, UV disinfection, and screw press.

Program Implementation Lessons Learned: [17]

The following observations are based on a survey of approximately 480 PG&E customers with a response rate of about 20% (99 respondents):

- During the last 10 years, 32 of 99 WWTF survey respondents received an audit, and PG&E supported 13 of the 32 audits. In the past 5 years, 42 out of 99 WWTF survey respondents had conducted energy efficiency projects; 28 of the 42 respondents who conducted projects used PG&E rebates (Table 1).
- WWTF operators were concerned about energy cost (68% of the respondents, or 67 respondents), but only slightly more than half of these respondents (36) addressed energy efficiency in new design or retrofit. Similarly, only 28 respondents used PG&E rebates, and only 33 respondents engaged in peak demand control (Table D-5).
- Energy efficiency projects favored variable frequency drives (61% of respondents), advanced instrumentation and control/SCADA systems (47% of respondents), dissolved oxygen sensors (34% of respondents), high-efficiency lighting (32% of respondents), and fine pore diffusers (29% of respondents) (Table D-6).

Table D-5. Summary of the Market-Related Results Addressing Energy Efficiency.

Table ES-1 – Summary of the Market-Related Results Addressing Energy Efficiency*		
	# of Plants	**Percentage**
Plants Engaged in Energy Efficiency Projects in Past Five Years	42	42%
Plants Received an Energy Audit in the Past Ten Years	32	32%
Audit Supported by PG&E	13	41%
Plants Greatly and Very Much Concerned About Energy Cost	67	68%
Plants Greatly and Very Much Concerned About Energy Efficiency Issues in New Design/Retrofit	36	36%
Plants which Used PG&E Rebates in their Projects	28	28%
Plants That Are Engaged in Controlling Peak Demand	33	33%

*Results are based on the responses from 99 filled-out questionnaires.

Source: PG&E, New Construction Energy Efficiency Program, *Energy Baseline Study for Municipal Wastewater Treatment Plants*, prepared by BASE Energy, Inc. (San Francisco, CA), http://www.pge.com/includes/docs/pdfs/mybusiness/energysavingsrebates/incentivesbyindustry/wastewater/wastewatertreatmentbaselinestudyreport.pdf, September 2006, Table ES-1.

[17] PG&E, New Construction Energy Efficiency Program, *Energy Baseline Study for Municipal Wastewater Treatment Plants*, prepared by BASE Energy, Inc. (San Francisco, CA), http://www.pge.com/includes/docs/pdfs/mybusiness/energysavingsrebates/incentivesbyindustry/wastewater/wastewatertreatmentbaselinestudyreport.pdf, September 2006.

Table D-6. Summary of the Major Results on WWTP Energy Efficient Technologies.

Table ES-3 – Summary of the Major Results on WWTP Energy Efficient Technologies*			
# of WWTP That Use Energy Efficient Technologies		*71 out of 99 (72%)*	
Energy Efficient Technologies Used		**# of Plants**	**Percentage of Total Responses**
Variable Frequency Drives (VFD)		60	61%
Application Where VFDs are Used	Pumps	55	56%
	Blowers	12	12%
	Compressors	3	3%
	Other	13	13%
Dissolved Oxygen (DO) Sensors		34	34%
Fine Pore Diffusers		29	29%
Advanced Instrumentation and Control / SCADA Systems		47	47%
High Efficiency Lighting		32	32%
Solar Aerators or Mixers		4	4%
High Efficiency Blowers		13	13%
Variable Intensity and/or Self-Cleaning Ultraviolet Lamps		4	4%
Pipe Internal Friction-Reducing Coating		9	9%
Screw Press for Sludge Dewatering		3	3%
Centrifuge for Sludge Dewatering		12	12%
Other Technologies		11	11%

*Results are based on the responses from 99 filled-out questionnaires.

Source: PG&E, New Construction Energy Efficiency Program, *Energy Baseline Study for Municipal Wastewater Treatment Plants*, prepared by BASE Energy, Inc. (San Francisco, CA), http://www.pge.com/includes/docs/pdfs/mybusiness/energysavingsrebates/incentivesbyindustry/wastewater/wastewatertreatmentbaselinestudyreport.pdf, September 2006, Table ES-3.

Wisconsin Focus on Energy: Water and Wastewater Program

Website: http://www.focusonenergy.com/Business/Industrial-Business/Wastewater.aspx

Program Overview: This program helps public and private wastewater systems "reduce energy use and operating costs".[18] The program uses energy advisors to: 1) conduct site audits to define and recommend energy efficiency opportunities and 2) develop a site-specific energy management plan. While conducting these tasks, the energy advisors may make the site operator aware of new energy-efficient technology options. Energy advisors also may help a participant through project development support (e.g., locating contractors, evaluating contractor proposals, and providing information and training resources that support project development).

Program Inception: 2000[19]

Mission: Provide wastewater systems with site energy audit and project development support that will facilitate their incorporating cost-effective energy efficiency options into existing, expanding, and new construction designs.

Program Metrics:[20] From 1999:

More than 140 GWh in potential electric savings

---

[18] Wisconsin Focus on Energy, Water and Wastewater website, http://www.focusonenergy.com/Business/Industrial-Business/Wastewater.aspx, February 2009.

[19] Kristi Kezar (SAIC), Joseph Cantwell (SAIC), "Lessons in Designing and Delivering Energy Efficiency Programs for the Water and Wastewater Industry", white paper, 2004, p. 2.

[20] ibid., p. 2.

Approximately 140 municipal and private participating facilities. (Wisconsin Department of Natural Resources has under its authority approximately 1,200 active industrial and municipal wastewater discharge permits and 1,180 potable water supply facilities.)

Program Implementation Lessons Learned:[21]

Design equipment and system alternatives that more closely match daily loads, especially early in a facility's 20-year life cycle, rather than following the existing practice of designing for 20-year peak load – Audits of wastewater systems in Wisconsin showed that the majority were designed and operated based on projected peak capacity over a forecasted 20-year facility life. The result is that wastewater systems often operated oversized or underloaded equipment relative to actual daily demand. Because facilities rarely operate at peak, especially in their early lives, authorities can achieve substantial energy savings through projects that install equipment sized to match usually experienced lower loads.

Respond to specific site needs with relevant services; recognize how facility capacity may affect choice of energy efficiency alternatives – At program inception, the model for promoting wastewater system energy efficiency projects called on the program representative to 1) perform an on-site survey; 2) present a survey report detailing potential energy-saving measures; 3) let the participant select projects from presented alternatives; 4) help the participant evaluate financing alternatives, resulting in extension of grant offers to the participant; and 5) monitor the project's progress through implementation.

The project identification and implementation support model changed as the program progressed to enable the program representative to use methods tailored to facilities of different sizes. One determination was that facility capacity and process complexity affect the types of energy efficiency measures that would be most cost-effective; thus, recognizing this relationship in choice of energy efficiency measures resulted in 1) an increase in the likelihood that a project would move forward to completion and 2) the attraction of a larger number of facilities, of different capacities, across the entire state, to the program because projects would be tailored to their specific characteristics. Once the program began to address specific site needs with relevant services, satisfied participants started to recommend the program to other wastewater system operators. These recommendations have led to increased program growth, including expansion of the program across the state.

For larger facilities, the main service provided by the program was simply the identification of measures and the preliminary assessment. Most facilities with capacity of 10 MGD or larger have engineers on staff or consulting firms with whom they work on a regular basis. In addition, many members of management were found to have technical backgrounds that provided them some knowledge of the energy-saving measures presented. Nevertheless, this group was found to lack the knowledge of innovative energy efficiency practices and technologies. Once the idea was "planted" and a preliminary level of analysis provided, they had the resources and ability to determine the benefits of the suggested project.

Another finding was that staff members in small and medium facilities (less than 10 MGD), especially those in rural areas, lack the time resources to identify appropriate energy-saving projects. However, they face the added dilemma of often having little or no in-house technical expertise, and typically they do not have an affiliation with a particular consulting firm.

In response to their needs, the program provided more detailed information on the recommended technology, as well as more explanation of the assessment methods and

---

[21] ibid., pp. 2–4. Program implementation lessons learned have been extracted from the referenced white paper, with rewording as necessary to present them as program best practices.

calculations. Feedback from participants in municipal facilities indicated that the detailed written report was helpful in explaining the nonmonetary benefits of suggested measures to managers and committees. In many cases, those making final decisions had no technical experience and based their opinions solely on financial considerations.

Set incentives to promote action – Facilities of all sizes, designs, and locations had one common barrier to implementing energy-saving measures – money. Municipal budgets were trimmed, leaving less money in equipment replacement funds and limiting the projects that facilities could pursue. Similarly, the industrial treatment facilities have to compete with manufacturing process improvements for capital funds. Even in facilities with healthy finances, the possibility of budget cuts decreased the likelihood that management would be willing to spend capital monies on energy efficiency projects, which are often viewed as being risky investments.

To help overcome the budget barrier, Focus on Energy provides cost-share grants to participants who commit to installing energy-efficient equipment. The program also has grant money available to offset the cost of feasibility studies to provide more detailed assessments of identified projects. These assessments can provide members of management with a reassuring second opinion that energy efficiency is a worthwhile investment or help them determine which equipment option will be most beneficial.

Aside from basic financial assistance, the grants are a strong indicator of program support for a project. Even a small investment on behalf of the program increased management confidence in a project. In several cases, a grant has been the deciding factor for facilities that were uneasy about investing in energy efficiency.

Work with the changing market – As the program progresses, the demand for services continues to grow. In an effort to respond to as many facilities as possible in a timely manner, the program has begun to distribute and encourage the use of self-evaluation tools and information. The main purpose of these materials is to provide a faster turnaround time for program participation at a lower cost to the program. The self-evaluation methods also allow facility staff to focus on projects that they feel deserves top priority.

Other program improvements include updated requirements for financial assistance and a faster application process with minimal paperwork. The program has begun to target industrial WWTFs across the state, such as those serving paper mills and food processing plants. Additionally, it is increasing its efforts to network facility staff and energy-efficient equipment and service providers in the area.

Continue to respond to participants' needs – The Water and Wastewater Program continues to evolve in ways that improve program offerings and increase the number of facilities benefiting from energy efficiency. Feedback from participants, trade groups, and others in the water and wastewater industry has been exceedingly positive. As energy costs continue to escalate and state and local governments struggle to cut costs without compromising services, the water and wastewater industry nationwide will provide energy-saving opportunities to help meet these future goals.

Influence future decisions – While encouraging the installation of energy-efficient equipment directly affects program participants' operating costs, the benefits of resource acquisition activities are guaranteed only for the lifetime of the equipment. To ensure that the program has long-lasting effects on energy efficiency choices in the water and wastewater industry, municipalities must offer services that influence future decisions.

Activities that transform market practices are also a critical part of any energy program. Providing education and training opportunities to explain the "whys" and "hows" of energy efficiency to operators allows facility staff to make informed decisions. Communication with regulatory agencies can influence future water and wastewater industry codes and design requirements to include energy efficiency and renewable technology considerations, such as CHP using biogas.

WASTEWATER UTILITY

Alabama
Montgomery Water Works & Sanitary Sewer Board

Alaska
Anchorage Water & Wastewater Utility

Arizona
Avondale, City of
Glendale, City of, Utilities Department
Mesa, City of
Peoria, City of
Phoenix Water Services Dept.
Pima County Wastewater Management
Safford, City of
Tempe, City of

Arkansas
Little Rock Wastewater Utility

California
Central Contra Costa Sanitary District
Corona, City of
Crestline Sanitation District
Delta Diablo Sanitation District
Dublin San Ramon Services District
East Bay Dischargers Authority
East Bay Municipal Utility District
El Dorado Irrigation District
Fairfield-Suisun Sewer District
Fresno Department of Public Utilities
Inland Empire Utilities Agency
Irvine Ranch Water District
Las Gallinas Valley Sanitary District
Las Virgenes Municipal Water District
Livermore, City of
Los Angeles, City of
Los Angeles County, Sanitation Districts of
Napa Sanitation District
Novato Sanitary District
Orange County Sanitation District
Palo Alto, City of
Riverside, City of
Sacramento Regional County Sanitation District
San Diego Metropolitan Wastewater Department, City of
San Francisco, City & County of
San Jose, City of
Santa Barbara, City of
Santa Cruz, City of
Santa Rosa, City of
South Bayside System Authority
South Coast Water District
South Orange County Wastewater Authority
South Tahoe Public Utility District
Stege Sanitary District
Sunnyvale, City of
Union Sanitary District
West Valley Sanitation District

Colorado
Aurora, City of
Boulder, City of
Greeley, City of
Littleton/Englewood Water Pollution Control Plant
Metro Wastewater Reclamation District, Denver

Connecticut
Greater New Haven WPCA
Stamford, City of

District of Columbia
District of Columbia Water & Sewer Authority

Florida
Broward, County of
Fort Lauderdale, City of
Jacksonville Electric Authority (JEA)
Miami-Dade Water & Sewer Authority
Orange County Utilities Department
Pinellas, County of
Reedy Creek Improvement District
Seminole County Environmental Services
St. Petersburg, City of
Tallahassee, City of
Toho Water Authority
West Palm Beach, City of

Georgia
Atlanta Department of Watershed Management
Augusta, City of
Clayton County Water Authority
Cobb County Water System
Columbus Water Works
Fulton County
Gwinnett County Department of Public Utilities
Savannah, City of

Hawaii
Honolulu, City & County of

Idaho
Boise, City of

Illinois
Decatur, Sanitary District of
Greater Peoria Sanitary District
Kankakee River Metropolitan Agency
Metropolitan Water Reclamation District of Greater Chicago
Wheaton Sanitary District

Indiana
Jeffersonville, City of

Iowa
Ames, City of
Cedar Rapids Wastewater Facility
Des Moines, City of
Iowa City

Kansas
Johnson County Wastewater
Unified Government of Wyandotte County/ Kansas City, City of

Kentucky
Louisville & Jefferson County Metropolitan Sewer District
Sanitation District No. 1

Louisiana
Sewerage & Water Board of New Orleans

Maine
Bangor, City of
Portland Water District

Maryland
Anne Arundel County Bureau of Utility Operations
Howard County Bureau of Utilities
Washington Suburban Sanitary Commission

Massachusetts
Boston Water & Sewer Commission
Massachusetts Water Resources Authority (MWRA)
Upper Blackstone Water Pollution Abatement District

Michigan
Ann Arbor, City of
Detroit, City of
Holland Board of Public Works
Saginaw, City of
Wayne County Department of Environment
Wyoming, City of

Minnesota
Rochester, City of
Western Lake Superior Sanitary District

Missouri
Independence, City of
Kansas City Missouri Water Services Department
Little Blue Valley Sewer District
Metropolitan St. Louis Sewer District

Nebraska
Lincoln Wastewater & Solid Waste System

Nevada
Henderson, City of
Las Vegas, City of
Reno, City of

New Jersey
Bergen County Utilities Authority
Ocean County Utilities Authority

New York
New York City Department of Environmental Protection

North Carolina
Charlotte/Mecklenburg Utilities
Durham, City of
Metropolitan Sewerage District of Buncombe County
Orange Water & Sewer Authority
University of North Carolina, Chapel Hill

Ohio
Akron, City of
Butler County Department of Environmental Services
Columbus, City of
Metropolitan Sewer District of Greater Cincinnati
Montgomery, County of
Northeast Ohio Regional Sewer District
Summit, County of

Oklahoma
Oklahoma City Water & Wastewater Utility Department
Tulsa, City of

Oregon
Albany, City of
Clean Water Services
Eugene, City of
Gresham, City of
Portland, City of Bureau of Environmental Services
Lake Oswego, City of
Oak Lodge Sanitary District
Water Environment Services

Pennsylvania
Hemlock Municipal Sewer Cooperative (HMSC)
Philadelphia, City of
University Area Joint Authority

South Carolina
Charleston Water System
Mount Pleasant Waterworks & Sewer Commission
Spartanburg Water

Tennessee
Cleveland Utilities
Murfreesboro Water & Sewer Department
Nashville Metro Water Services

Texas
Austin, City of
Dallas Water Utilities
Denton, City of
El Paso Water Utilities

Fort Worth, City of
Houston, City of
San Antonio Water System
Trinity River Authority

Utah
Salt Lake City Corporation

Virginia
Alexandria Sanitation Authority
Arlington, County of
Fairfax County
Hampton Roads Sanitation District
Hanover, County of
Henrico, County of
Hopewell Regional Wastewater Treatment Facility
Loudoun Water
Lynchburg Regional Wastewater Treatment Plant
Prince William County Service Authority
Richmond, City of
Rivanna Water & Sewer Authority

Washington
Everett, City of
King County Department of Natural Resources
Seattle Public Utilities
Sunnyside, Port of
Yakima, City of

Wisconsin
Green Bay Metro Sewerage District
Kenosha Water Utility
Madison Metropolitan Sewerage District
Milwaukee Metropolitan Sewerage District
Racine, City of
Sheboygan Regional Wastewater Treatment
Wausau Water Works

Water Services Association of Australia
ACTEW Corporation
Barwon Water
Central Highlands Water
City West Water
Coliban Water Corporation
Cradle Mountain Water
Gippsland Water
Gladstone Area Water Board
Gold Coast Water
Gosford City Council
Hunter Water Corporation
Logan Water
Melbourne Water
Onstream
Power & Water Corporation
SEQ Water
South Australia Water Corporation
South East Water Limited
Sunshine Coast Water
Sydney Catchment Authority
Sydney Water
Wannon Regional Water Corporation
Watercare Services Limited (NZ)
Water Corporation
Water Distribution Brisbane City Council
Western Water
Yarra Valley Water

Canada
Edmonton, City of/Edmonton Waste Management Centre of Excellence
Lethbridge, City of
Regina, City of, Saskatchewan
Toronto, City of, Ontario
Winnipeg, City of, Manitoba

STORMWATER UTILITY

California
Fresno Metropolitan Flood Control District
Los Angeles, City of, Department of Public Works
Monterey, City of
San Francisco, City & County of
Santa Rosa, City of
Sunnyvale, City of

Colorado
Aurora, City of
Boulder, City of

Florida
Orlando, City of

Iowa
Cedar Rapids Wastewater Facility
Des Moines, City of

Kansas
Lenexa, City of
Overland Park, City of

Kentucky
Louisville & Jefferson County Metropolitan Sewer District

Maine
Portland Water District

North Carolina
Charlotte, City of, Stormwater Services

Pennsylvania
Philadelphia, City of

Tennessee
Chattanooga Stormwater Management

Texas
Harris County Flood Control District, Texas

Washington
Bellevue Utilities Department
Seattle Public Utilities

STATE

Connecticut Department of Environmental Protection
Kansas Department of Health & Environment
New England Interstate Water Pollution Control Commission (NEIWPCC)
Ohio Environmental Protection Agency
Ohio River Valley Sanitation Commission
Urban Drainage & Flood Control District, CO

CORPORATE

ADS LLC
Advanced Data Mining International
AECOM
Alan Plummer & Associates
Alpine Technology Inc.
Aqua-Aerobic Systems Inc.
Aquateam–Norwegian Water Technology Centre A/S
ARCADIS
Associated Engineering
Bernardin Lochmueller & Associates
Black & Veatch
Blue Water Technologies, Inc.
Brown & Caldwell
Burgess & Niple, Ltd.
Burns & McDonnell
CABE Associates Inc.
The Cadmus Group
Camp Dresser & McKee Inc.
Carollo Engineers Inc.
Carpenter Environmental Associates Inc.
CET Engineering Services
CH2M HILL
CRA Infrastructure & Engineering
CONTECH Stormwater Solutions
D&B/Guarino Engineers, LLC
Damon S. Williams Associates, LLC
Ecovation
EMA Inc.
Environmental Operating Solutions, Inc.
Environ International Corporation
Fay, Spofford, & Thorndike Inc.
Freese & Nichols, Inc.
ftn Associates Inc.
Gannett Fleming Inc.
Garden & Associates, Ltd.
Geosyntec Consultants
GHD, Inc.
Greeley and Hansen LLC
Hazen & Sawyer, P.C.
HDR Engineering Inc.
HNTB Corporation
Hydromantis Inc.
HydroQual Inc.
Infilco Degremont Inc.
Jason Consultants LLC Inc.
Jordan, Jones, & Goulding Inc.
KCI Technologies Inc.
Kelly & Weaver, P.C.
Kennedy/Jenks Consultants
Larry Walker Associates
LimnoTech Inc.
Lombardo Associates, Inc.
The Low Impact Development Center Inc.
Malcolm Pirnie Inc.
Material Matters, Inc.
McKim & Creed
MWH
NTL Alaska, Inc.
O'Brien & Gere Engineers Inc.
Odor & Corrosion Technology Consultants Inc.
Parametrix Inc.
Parsons
Post, Buckley, Schuh & Jernigan
Praxair, Inc.
RMC Water & Environment
Ross & Associates Ltd.
SAIC
Siemens Water Technologies
The Soap & Detergent Association
Smith & Loveless, Inc.
Southeast Environmental Engineering, LLC
Stearns & Wheler, LLC
Stone Environmental Inc.
Stratus Consulting Inc.
Synagro Technologies Inc.
Tetra Tech Inc.
Trojan Technologies Inc.
Trussell Technologies, Inc.
URS Corporation
Wallingford Software
Westin Engineering Inc.
Wright Water Engineers
Zoeller Pump Company

INDUSTRY

American Electric Power
American Water
Anglian Water Services, Ltd.
Chevron Energy Technology
The Coca-Cola Company
Dow Chemical Company
DuPont Company
Eastman Chemical Company
Eli Lilly & Company
InsinkErator
Johnson & Johnson
Merck & Company Inc.
Procter & Gamble Company
Suez Environnment
United Utilities North West (UUNW)
United Water Services LLC
Veolia Water North America

Note: List as of 11/10/09

IWA Publishing's authorised EU representative for General Product Safety Regulations is Diane D'Arras, 15 rue Duret, 75116 Paris, France, e-mail: safety@iwap.co.uk.

Printed and bound by CPI Group (UK) Ltd, Croydon, CR0 4YY
17/09/2026
02218799-0013